Monitoring & Evaluation: Data Management Systems

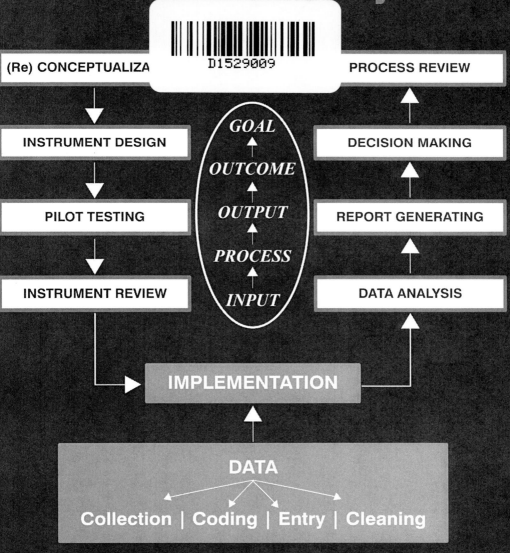

Bongs Lainjo, MASc, Engineering
Former UN Senior Advisor

Monitoring & Evaluation: Data Management Systems

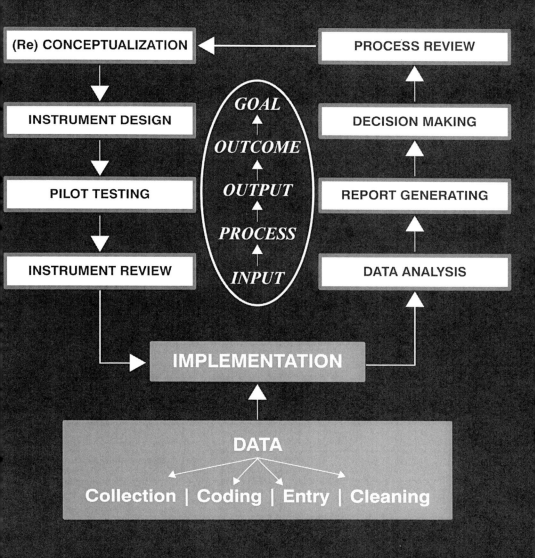

Bongs Lainjo, MASc, Engineering
Former UN Senior Advisor

Copyright © 2015 by Bongs Lainjo

All rights reserved. No part of this book may be reproduced or transmitted in any form or by any means, electronic or mechanical, including photocopying, recording, or any information storage and retrieval system, without permission in writing from the author.

ISBN: 978-0-9909778-2-7
Library of Congress Control Number: 2015905370

10 9 8 7 6 5 4 3 2 040315

Excerpt From: Bongs Lainjo. "Monitoring and Evaluation: Data Management Systems." eBook Edition.

∞This paper meets the requirements of ANSI/NISO Z39.48-1992 (Permanence of Paper)

Table of Contents

FOREWORD ix
PREFACE xiii
ABBREVIATIONS xvii
1. Introduction 1
   1.1 Program Life Cycle (PLC) 3
2. Background to the Literature Review 5
3. Objectives 6
4. Literature Review 6
   4.1 Paris Declaration/Core Principles of Managing
       for Development Results (MfDR) Conundrum 9
   4.2 Research Process 11
   4.3 Analysis of Data 13
   4.4 Analyzing Information 15
   4.5 Development Assistance Committee Criteria 16
5. Methodology 17
6. Target Audience 17
7. Results-Based Management 18
8. Monitoring and Evaluation (M&E) 19
   8.1 Basic Principles of Monitoring and Evaluation 19
   8.2 Monitoring and Evaluation Plan 21
   8.3 Program Indicator Screening Matric (PRISM) 25
       8.3.1 Introduction 25
       8.3.2 General Objectives 26
       8.3.3 Specific Objectives 26
       8.3.4 Relevance of the PRISM 27
       8.3.5 Audience Targeted by the PRISM 27
       8.3.6 Evaluation Life Cycle 27
       8.3.7 Program Design Frameworks 30
       8.3.8 Logframe 31
           8.2.8.1 M and E – Analysis Case Study 33
       8.3.9 The Matrix 36
          In short, the matrix should
          contain the following: 37
       8.3.10 Themes of the PRISM 39
       8.3.11 Criteria of the PRISM 39

- 8.3.12 Implementation of the PRISM 41
- 8.3.13 Algorithm of the PRISM 41
- 9. Application of Data Management Framework 43
  - 9.1 Principles of Data Management Framework 43
  - 9.2 Variables and Indicators 44
    - 9.2.1 Variables 44
  - 9.3 Basis Statistics 46
  - 9.4 Sampling 52
  - 9.5 Measurement Scales 55
    - 9.5.1 The Importance of Choosing the Appropriate Measurement Scale 57
  - 9.6 Data and Information 58
    - 9.6.1 Organization of the Data 58
  - 9.7 Statistical Software Packages 59
    - 9.7.1 Epi Info 60
    - 9.7.2 SPSS 60
    - 9.7.3 SAS 60
    - 9.7.4 STATA 61
- 10. Student Test 62
  - 10.1 Survey Data Analysis in STATA 63
  - 10.2 Analyzing Survey Data in STATA 64
    - 10.2.1 Simple Random Sampling 65
    - 10.2.2 Stratified Sampling 66
    - 10.2.3 Cluster Sampling 66
    - 10.2.4 Stratified Cluster Sampling 67
    - 10.2.5 ANOVA in Stata 72
    - 10.2.6 Real Life Case Study (t-test) 74
      - 10.2.6.1 Study Design 74
      - 10.6.2.2 Survey Methodology 75
- 11. Data Management Model 76
  - 11.1 Conceptual Framework 77
    - 11.1.1 Conceptualization 78
    - 11.1.2 Instrument Design 79
    - 11.1.3 Pilot Testing 80
    - 11.1.4 Implementation 81
    - 11.1.5 Data Collection 81
  - 11.2 Teams 82
  - 11.3 Data Cleaning and Consolidation 83
  - 11.4 Data Analysis 93

11.5 Report Writing 101
11.6 Decision Making 101
Conclusion 103

APPENDIX 105
1: Data Management Framework 105
2a: Program Indicator Screening Matrix (PRISM) 106
2b: PRISM Case-Study – Intra-Group Screening 107
2c: PRISM Case-Study – Intra-Group Screening 108
2d: PRISM Case-Study – Inter-Group Screening 109
2e: PRISM Case-Study – Inter-Group Screening 110
3: DAC 2010 Glossary of Key Terms in M and E 111
Bibliography 125
Index 131
M and E: Data Management Systems 135

LIST OF FIGURES
Figure 1: Program Life Cycle Layout 4
Figure 2: Self Assessment Spider Chart 7
Figure 3: Paris Agreement Framework 9
    Figure 3.1: Highlight of the Paris agreement protocol 10
    Figure 3.2: Criteria For Analyzing M&E 15
Figure 4: Evaluation Life Cycle 28
Figure 5: M&E Cycle 29
Figure 6: Logframe 31
Figure 7: Mean Distribution Of Staff Sick Leave From
    2008 To 2010 35
Figure 8: Algorithm Of The PRISM 41
Figure 9: Frequency Graph 47
Figure 10: Monthly Mean Number of Oral
    Contraceptive Clients 76
Figure 11: Data Management Life Cycle (Framework) 77
Figure 12: Percentage of Randomized Health Facilities Using
    Graph for Monitoring Services. 94
        Figure 12a: Port Orly Health Facility 99
        Figure 12b: Cross-Section of Port Orly Health
            Facility Ante-Natal Clients 100

LIST OF TABLES

Table 1: Program Design Frameworks 30
Table 2: Strategic Framework and M & E Summary Table 32
Table 3: OECD/DAC Working Group Criteria Definition 33
Table 4: Program Indicator Screening Matrix (PRISM) 38
Table 5: Frequency table 46
Table 6: Analytic Capabilities Of Four Statistical Software
          Packages to Analyze Complex Survey Data 61
Table 7: Expenditure 86
Table 8: Education Levels 90
Table 9: Households 91
Table 10: Expenditure (Cleaned) 92
Table 11: Health Service Providers 95

Foreword

Bongs Lainjo's "M and E: Data Management Systems" is a document I really could have used when, back in 1966, I took my first overseas position in health and development! The need was there. I even knew it, but the body of experience that would ultimately give rise to this overview of tools that was needed to help with the monitoring and evaluation of programs had not yet been developed. Like most young advisors, I learned by trial and error, applying the useful, but not always the "real-world" academic training to the realities of working in Asian, and later African, settings.

Since then, the fields of program development and management, and within this the area of program monitoring as a major aspect of evaluation and guidance, have advanced enormously. This is due not just to the accumulation of experience over time in ever-widening circles of programs and projects in more and varied country settings, but also to the development of tools to assist in systematic planning and evaluation. Today, we find a proliferation of reports and analyses, which provide a rich base for the field investigator to tap into, for deciding how best to design and manage projects, and how best to design the types of monitoring and evaluation components that will yield the most helpful information for guiding his/her project. Also, now available are powerful statistical tools that can help turn raw data into meaningful pictures of what is actually being accomplished, relative to program objectives.

This manual can be considered a distillation of the collected experiences represented by the accumulation of literature on data management systems, as well as of the monitoring and evaluation examples that lie at the heart of project management. This text provides a "primer" for fieldworkers engaged in efforts to improve their projects and programs. In other words, this is that volume that would have been so helpful to me, had I had it (or its 1966 equivalent) available to me when I first entered the field! This is the resource that would have sat on my desk in my office, and would have accompanied me into the field, to help me determine how best to proceed with the projects I was involved in, in the many countries over the next decades. Indeed, it is not only the neophyte or novice who will be able to benefit from this text.

Bongs Lainjo, MASc, Engineering

Bongs Lainjo has produced a book that caters to a range of audiences starting with different skill and experience levels, from the student first trying to gain an appreciation of what "management" involves in health and development programs, to the experienced field practitioner, trying to determine specific M & E approaches to apply to his current project. In preparing this resource, Mr. Lainjo has used his own decades of field experience in "developmental" settings, combined with his knowledge of the technological improvements in statistical analysis to guide the user in addressing his own program's needs. He does this by presenting a framework for understanding the organizational matrix for the program or project under review, with a particular focus on the needs of the data management system for selecting and collecting appropriate data, and then analyzing it in ways that can enhance ongoing decision-making.

From this logical framework, the author takes the reader through a review of the basic principles of monitoring and evaluation, and what is required in planning for their usage. He also notes that, in part, based on his own background of often working on projects supported in large part with external funding, the perspectives of "donors", including often the requirement that the data generated needs to be available to wider audiences, must also be recognized (donors also have an annoying fondness for "accountability"!). The importance of thinking through "team" involvement in evaluation, and its feedback to program decision-making, is also addressed.

Mr. Lainjo introduces the reader to "PRISM", the acronym representing the Program Indicator Screening Matrix, and walks us through it's components, before moving on to the application of the data management framework. In this, he provides the reader with an overview of the key concepts and how these fit into the foregoing framework for assessing programs. This includes a basic review of what is meant by variables and indicators, and what is needed to operationalize these, as well, he reviews basic sampling issues and the appropriate measurement scales. While for some readers, this will simply be a useful review, for others it may be their first real introduction to systematic "evaluation thinking".

Next, the author turns to the question of how to organize the program data in order for it to be analyzed meaningfully. In this section, he discusses briefly some of the more readily available statistical software packages, which may be appropriate in given settings, with some guidance on how to make this determination. Along the way, he alerts the reader that the reader/user is the one, who must ultimately decide, on the basis of the conceptual and technological tools available to him, and how best to benefit from "monitoring and evaluation"! While "the typical management system" is his main focus, the author recognizes that "one size does not fit all". This manual will also help the reader to determine what best fits his own and his program's needs!

Finally, on a personal note, I have been aware of Bongs Lainjo's work for several decades. I first became aware of him in the mid 1980's, through a CDC colleague who worked closely with him on the USAID "DISH" project in Africa who only spoke in very complimentary terms of Bongs' work on that project. This colleague and I then had the opportunity to work with Bongs directly at our CDC offices in Atlanta. During that period, Bongs continued to demonstrate a very high degree of professionalism, while also interacting constructively with other colleagues in moving these activities forward. Since that time, he and I have maintained contact both at a personal as well as on a professional level. Along the way, I have reviewed and discussed with him some of his peer reviewed papers that he presented at International conferences. It has turned out to be a satisfying collaboration. Having also participated in reviewing this Primer, I consider this as a further extension of that collaborative effort.

J. Timothy Johnson, MSc., DrPH

PREFACE

Over the past decades, international funding agencies, as well as, recipient national governments in developing countries have expended substantial amounts of money on developing and implementing programs. Both have increasingly been interested in putting in place sustainable and effective programs. This has made it necessary for all the players involved in a donor funded program to understand the dynamics of these programs. This primer is intended to help bridge the gap that currently exists between different players - donors as supply and beneficiaries as demand parties – who are both involved in program management. This is because program management has been globalized to include the different players with diverse experiences and varying levels of exposure.

The manual is an accumulation of over three decades of experience achieved through working in academia and different developing countries. I have also benefitted from this experience by listening to my national counterparts who are working diligently in the government and who also provide meaningful and sensitive feedback. Such approaches not only helped me facilitate my collaborative efforts with government colleagues, but also gave me an opportunity to establish a mutual degree of confidence and trust in each other. Above all, I also had the opportunity to redirect my intervention strategy. The high degree of trust enabled to effectively establish a continuous and effective participatory process among counterparts which yielded significant and sustainable results. For example, I was able to train nationals and work with them in refining program indicators: developing and pre-testing questionnaires, conducting household surveys and censuses, conducting community health surveys, processing data and analyzing them, and writing reports.

This manual, which contains Monitoring and Evaluation (M&E) and Data Management (DM) sections, is presented and directed to a diverse group of audiences (from the neophyte to the sophisticated). A subsection of Basic Statistics is also included in order to help the reader understand most of the rudimentary statistical concepts that are necessary when carrying out or going through a research project or analyzing data. It is my belief that this text will furnish the reader with the necessary knowledge that is required for them to execute a research project, and most specifically in the area of data management. This text also includes, to a greater extent, primary data with occasional cases of secondary data.

A number of case studies have been given to help the reader comprehend the issue that I believe required more emphasis. In that regard, I have developed all my case studies from real life circumstances. They have also been included because they are not only fun and interesting, but they also go much further to help readers find out valuable and useful information. They represent another attempt to demystify statistics. Thus, as users of the primer, we need to constantly remind ourselves that these are simply numbers driven by rules and assumptions..We must realize that we inaccurately characterize them as lies, damned lies and statistics simply because we fail to follow and understand the assumptions that govern statistics. As it turns out, most of what we do is driven by statistics: time to wake up in the morning, time to have breakfast, time to go to work, time for lunch, the price we pay at the store for different commodities etc. These are all statistics. This manual or primer, if you will, is unique, thorough, inclusive and practical!

It would be inappropriate, therefore, and unethical for me to fail to acknowledge with gratitude the many contributions of the various people who have made the completion of this primer successful. This project could never have materialized, without direct and indirect contributions from colleagues and from the former organizations I worked for. I am very grateful to the UN, USAID and Columbia University. They did not only give me an opportunity to serve in under privileged communities, but they also allowed me the opportunity to live with them and learn and appreciate different cultures. This was an experience that I could never have learned from any text book. Their perpetual support was indeed an inspiration by itself.

My profound gratitude also goes to P.M. "Jesse" Brandt, MA, and former UNFPA Senior Advisor whose relentless and unyielding efforts and encouragement were very significant motivating factors in this initiative. Mr. Brandt has extensive experience working specifically in Asia, where he also worked as an advisor for WHO as a member of the multidisciplinary Country Support Team for South, East and Central Asia first headed by UNICEF and then by UNFPA Directors and headquartered in Kathmandu, Nepal.

My appreciation also goes to Dr. Tim Johnson, whose international health experience in 40 countries mostly in Africa and Asia, combined with his background as an academic researcher and then as director of international reproductive health programs at the US Centers for Disease Control Atlanta, provided me with a useful sounding board that invaluably moved this initiative one level higher. Dr. Johnson's comments during the review process contributed substantially to mitigating the technical jargon that authors of similar manuals have the tendency of using. This helped make the primer simpler and reader-friendly. Finally, this primer is dedicated to my parents – my mother, Cecilia Banboye my mom, who taught me the multiplication tables and, also, my father VT Lainjo, who taught us to always stand up and be counted-- for what we believe in.

Bongs Lainjo,
MASc, Engineering

ABBREVIATIONS

ALNAP - The Active Learning Network for Accountability and Performance

DAC - Development Assistance Committee

DM - Data management

HELP - Humanitarian Evaluation and Learning Portal

M&E - Monitoring and Evaluation

OECD - The Organization for Economic Co-operation and Development

USAID - United States Agency for International Development

PRISM - Program Indicator Screening Matrix

UN – United Nations

CIDA – Canadian International Development Agency

EU – European Union

AusAID – the Australian Agency for International Development

DfID – Department for International Development

WB – World Bank

WFP – World Food Program

1. Introduction:

Data Management is a concept whose time has come. It has evolved gradually and surely from a limited number of fields to its current state where its ubiquity can no longer be shrouded by any forces. Its cross-sectional appeal has made it popular in fields ranging from science, industry, business, government to organizations -- just to mention a few. This diversity has generated its own controversies. For example, there are as many interpretations and understandings of the word 'Data Management' as there are interested stakeholders. The silver lining among these diversities is the commonality of its being able to generate and transform data for informed decision-making processes.

When one looks at the myriad fields involved in managing data, one is struck by the simplicity on the one hand and complexity on the other. For instance, the store clerk who tracks and manages commodity supplies is interested in how stock flows and at the same time tries to minimize stock-outs. On the other spectrum, we have the rocket scientist who is challenged by complex systems dependent on reliable and stable data sets that are required in order to enable the successful accomplishment of certain goals and objectives.

Data Management as we know it today has also been popularized by the strong demands on companies and organizations to produce meaningful results. This is more pronounced in the area of Monitoring and Evaluation (M and E). In general and where available and used judiciously, M and E activities have served as the driving force that underscores the importance and usefulness of data management.

From an M and E perspective, data management will be defined as a complementary force towards the intermediate output, outcome and goal/strategic objectives. In program management, output, outcome and goals are all strategic elements of a logical framework (log frame), which are, generally, used by governments and implementing program agencies. The others, intermediate results and strategic objectives are primarily used by the US government. The United States Agency for International Development (USAID) is a strong proponent of this framework. From a program management view, the outputs (or lower level indicators) are generally monitored while the higher level indicators (outcome and goals) are generally evaluated. In the former case, the objective is to assess the degree of progress (or lack of it) towards achieving the planned results. In the latter case, the objective is to establish the level of satisfaction among the beneficiaries of the intervention. There are two distinct categories of this that are generally recognized by program implementing partners – process and impact.

Although there are different perspectives from which data management (DM) in health and development programs and projects can be approached, we will approach it broadly as encompassing a system of components that range from the conceptual stage to the decision making level. Details of this definition will be presented elsewhere as a model (see Appendix 1) as well as below. With this generalized framework in mind, this manual will focus particularly on the sorts of practical problems encountered in the field that provide policy makers, managers, and ultimately front-line field workers, with the information needed to make better, data-based, decisions. The major monitoring and evaluation (M&E) skills, and particularly relevant tools, will be the specific focus of this manual.

The framework presented below is envisaged primarily for managing primary data sets derived from the program or project under review; although in certain cases, it also applies to secondary data sets obtained from other sources, but are seen as providing relevant information. Further discussion and details on primary and secondary data will be presented elsewhere in the manual.

One may first ask what the target audience is, and is there more than one audience to whom this manual would prove particularly useful? As will become clear in subsequent sections, there are several audiences, and indeed a list of potential target audiences is included later. While at the broadest level, any 'field practitioner', from upper level policy makers and managers to the lower-level 'field implementers' can gain insights from this distillation of the author's practical and academic experience in M&E. However, the ones to whom this volume is most directed are front-line staff, who already have a good sense of what the objectives of their programs are, but they need better tools, and better conceptual frameworks to make sense of their programs, through the improved management of program data, along with such ancillary tasks as the generation and collection of relevant data to their programs.

With these target audiences in mind, the author has made great efforts to present over three decades of his experience in a concise, logical, easy to follow and reader friendly way, with illustrative examples. He has also tried to make it as free as possible of technical jargon so that it may even benefit the layman. Thus, there is ample opportunity for different categories of user groups ranging from the uninitiated neophyte to the most sophisticated guru, or for the academic or trainer helping to guide the neophyte to gain the knowledge to manage data better. Even though it is assumed that any user of this manual will start with at least some basic arithmetic skills, and some general academic or experience-based sense of how programs work.

It is the author's hope, as well, that this primer will also serve as an effective and practical reference manual that will contribute in field operations, while also filling certain gaps that M and E specialists are currently faced with. Further to this, attempts have also been made to make the document as comprehensive as possible. The method thus, involves providing the user with a 'one stop shop', illustrates ideas by making available the relevant case studies, gives simple and clearly defined models, and above all outlines various levels of actual field experience.

1.1 Program Life Cycle (PLC)

From a program management perspective, interventions are characterized as either projects or programs. In each case, there are shared communalities. Figure 1 gives a graphic representation of the different components considered in many cases during development initiatives. These components include:

Conception: A stage during which an idea is "born", generally, based on existing development landscapes or as a continuation of a previously accomplished plan.

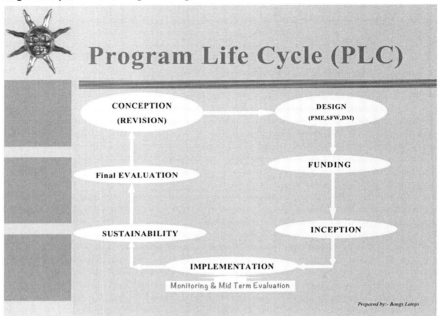

Figure 1: Program Life Cycle Layout

Design: This stage applies to the transformation of the conceived idea into a blueprint. It is during this process that proposals are developed, revised and finalized before being officially circulated.

Funding: The survival of any proposal depends on the level of available funding. It is at this level when contractors lobby and advocate with funding agencies or donors for possible funding. It is a very competitive process as each contractor needs to stand up and convince funding agencies that their proposal is more practical, more cost effective, achievable and relevant to the intended beneficiaries. It is also during this stage that contractors collaborate in some circumstances with their competitors for partnerships.

Inception: After a contract has been awarded, the successful contractor(s) get together with the potential donors and layout a future "way forward" strategy. This is a detailed and revised plan with a more refined M and E plan, work plans, the relevant stakeholders, implementing partners and, of course, with the role of the beneficiary national government.

Implementation: The above inception report serves as a road map of implementation. Different parties are, at this stage, engaged in clarifying levels of effort and responsibilities. As well, it is at this stage that developed monitoring plans are finalized through a consultative process. The original strategic frameworks are also reviewed and revised to better reflect the environmental dynamics that may have changed the development landscape between the program design and its implementation. It is also during this stage, that evaluation exercises are conducted. Some donors require a single evaluation during the project life time; others insist on conducting a mid-term and an end term evaluation.

Sustainability: This stage as outlined in the manual remains quite ambiguous, controversial and divisive. Depending on who is involved, there are different definitions associated with this development stage. However, it is at this level that evaluation teams are required as part of their terms of reference to establish the degree of sustainability of a given project. Quite frequently (and mostly), this evaluation exercise is conducted soon after project completion. This makes the evaluation process even more cumbersome given the copious number of indicators involved and the likelihood of most of them losing their effectiveness during an unknown period of time.

Finally, this figure also helps establish some common ground among readers. It is meant so that readers could better situate their concept or programs more objectively. It is meant to clarify certain confusing perceptions that some may have about programs and projects. It is also fair to state here that the figure varies from stakeholder to stakeholder. Hence this is simply one of several possibilities.

2. Background to the Literature Review

This document consists of a summary as well as a review of the important themes that relate to any data management system. It aims to help minimize the gap that currently exists in the area of data management by presenting these themes in a precise non-technical language, which will go a long way in generating interest among readers from different spheres of life. The groups which may find this document particularly useful include staff of academic institutions, government administrators, Non-Governmental Organizations (NGOs) and Program Implementing Partners, such as the staff of organizations nominated as Implementing Partners for USAID-funded programs.

3. Objectives

i. *General objective:*

- To describe and elaborate on the various basic concepts of data management systems.

ii. *Specific objectives:*

- To describe and illustrate a typical integrated data management system.
- To describe and differentiate among data management frameworks and their components.
- To introduce descriptive analysis.

4. Literature Review

M&E is something people do in their daily lives (Ramboll, 2005). They observe what is happening or has happened around them – for example, they observe the amount of money they have in their bank account at a specific time of the month. They assess the current situation and then compare it to their expectations or goals. If there is a difference; they decide whether the difference is significant – did they save the amount they had planned towards their annual holiday, and if not, will that significantly affect their holiday plans. They then consider ways in which they can address these differences or shortcomings.

M&E can be considered as a practical management tool which is used in reviewing performance (Ramboll, 2005). M&E helps in learning from experience, which can consequently be used to improve the designing as well as the functioning of projects. Quality assurance and accountability are integral components of M&E, which help to ensure that project objectives are met, in addition to achieving key outputs and impacts (Ramboll, 2005).

Diakonia developed a self-assessment tool in Cambodia in the early 2000s, which included eight capacity areas. Diakonia's tool uses a scoring system which is similar to those used by USAID and Pact.

Monitoring and Evaluation was one of the key capacity areas. The full list of the capacity areas include:

i. Advocacy problem identification
ii. Research
iii. Goal setting
iv. Indicators
v. Stakeholder analysis
vi. Action plan
vii. Coalition building
viii. Monitoring & evaluation

In order to make the strengths as well as the weaknesses of an organization's capacity easy to identify, a visual representation of the model is provided. It uses a spider chart as shown in figure 2 below. According to Raynor (2009), a critic of such a model, it is too skills-oriented; although, it covers all the skill areas of interest.

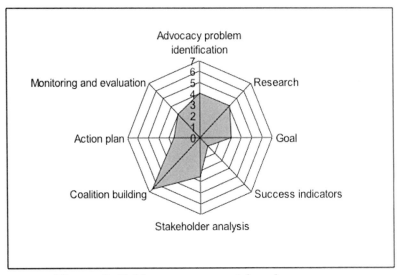

Figure 2: Self assessment spider chart

According to de Mendoza (2010), all donors are expected to follow monitoring processes regularly. This, in some instances, includes a six month review with both the stakeholders and beneficiaries. The aim is to help strengthen the self assessments of progress, improve documentation of the implementation processes and facilitate timely modifications as needed (de Mendoza, 2010).

Ramboll (2005) observed that it is necessary to specify the purpose as well as the scope of the M&E because it helps clarify what is to be expected of the M&E procedure, how comprehensive it ought to be and what time and resources will be needed to implement it.

An M&E Plan helps identify the report, which includes the M&E information and sets out the forums or meetings at which the information or the reports will be presented and discussed (WFP, 2014). Thus, the M&E Plan sets out the important formal feedback opportunities as well as ensures that M&E reports are made available to all stakeholders and that the appropriate formal as well as informal discussions are held concerning indispensable findings (WFP, 2014).

According to WFP (2014), the following guidelines should be followed when writing a Monitoring and Evaluation report:

1. There should be consistency with regard to the amount of information to be conveyed.
2. Focus should be on the results achieved rather than on the expected results.
3. A section must be included describing the reasons for the collection of data.
4. A section must be included describing the sources of data and the methods used in collecting data for the findings to be objectively verifiable.
5. It is essential to be clear on who the target audience is.
6. The language used in writing should be understandable to the target audience.
7. The progress reports should be submitted on time.
8. A brief summary, about a page, should be provided at the beginning which accurately captures the content as well as the recommendations in the report.
9. Any technical terns or acronyms should be defined. There should also be consistency in the use of definitions and terminologies.
10. Complex data should be presented with the help of summary tables, figures, graphs, maps and photographs.

11. Only the most significant words or key points should be highlighted using stresses such as bold or italics.
12. References for sources as well as authorities should be included.
13. A table of contents should be included for reports with more than five (5) pages.

4.1 Paris Declaration and Core Principles of Managing for Development Results (MfDR) Conundrum

Originally called "The Aid Effectiveness Pyramid" (2005), the Paris Declaration Pyramid was supported by over 100 donors and developing countries (OECD 2005). The Paris Declaration is touted as being concerned with changing behavior under the premise that increased aid flows are unlikely to make a serious dent in global poverty; if donors do not change the way they go about providing aid and developing countries do not enhance the way they currently manage it (OECD 2006). The donors and recipients in Paris agreed to adopt the strategic framework known as MfDR, or managing for development results. MfDR consists of broad strategic planning and risk assessment/management exercises, stated progress monitoring objectives and (nonspecific) outcome evaluation mechanisms.

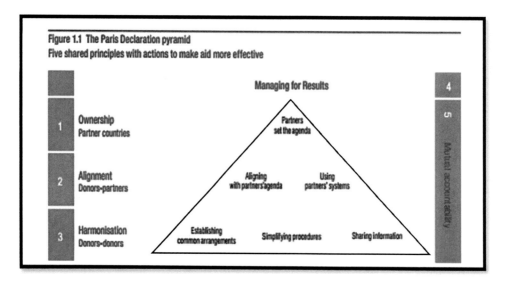

Figure 3 represents the Paris agreement framework in a nutshell.

PARIS DECLARATION INDICATORS

OWNERSHIP

1. Countries put in place national development strategies with clear strategic priorities.

ALIGNMENT

2. Countries develop reliable national fiduciary systems or reform programmes to achieve them.
3. Donors align their aid with national priorities and provide the information needed for it to be included in national budgets.
4. Co-ordinated programmes aligned with national development strategies provide support for capacity development.
5a. As their first option, donors use fiduciary systems that already exist in recipient countries.
5b. As their first option, donors use procurement systems that already exist in recipient countries.
6. Country structures are used to implement aid programmes rather than parallel structures created by donors.
7. Aid is released according to agreed schedules.
8. Bilateral aid is not tied to services supplied by the donor.

HARMONISATION

9. Aid is provided through harmonised programmes co-ordinated among donors.
10a. Donors conduct their field missions together with recipient countries.
10b. Donors conduct their country analytical work together with recipient countries

MANAGING FOR RESULTS

11. Countries have transparent, measurable assessment frameworks to measure progress and assess results.

MUTUAL ACCOUNTABILITY

12. Regular reviews assess progress in implementing aid commitments.

Figure 3.1 is a comprehensive highlight of the Paris agreement protocol.

Upon closer inspection, no implementing guidelines have been found that might show how a developing country could possibly be aligned with the model. It appears as if the model is meant to guide the thinking of the donors, not the beneficiaries. For example, in speaking of common arrangements, someone might ask "What is meant by 'common arrangements' other than the existing development process, which might not have been successful in the past? Similarly, one might wonder whether there are specific 'simplified procedures' to follow other than those a country currently applies and uses which might be complicated and obstructive to the development process?

> **Box 2: Definitions and Principles of Managing for Development Results (MfDR)**
>
> **Definitions**
>
> Managing for development results (MfDR) is a management strategy focused on development performance and on sustainable improvements in country outcomes. It provides a coherent framework for development effectiveness in which performance information is used for improved decision-making, and it includes practical tools for strategic planning, risk management, progress monitoring, and outcome evaluation.
>
> **MfDR Core Principles**
> 1) Focus the dialogue on results at all phases.
> 2) Align actual programming, monitoring, and evaluation activities with the agreed expected results.
> 3) Keep the results reporting system as simple, cost-effective, and user-friendly as possible.
> 4) Manage *for*, not *by*, results.
> 5) Use results information for management learning and decision making, as well as for reporting and accountability.

What happened to the Paris Declaration? Over the six or seven years since 2005, the results of three subsequent surveys undertaken to understand the process and results of the implementation of MfDR, showed that only one out of 12 goals was achieved (OECD 2011). This was the one wherein donor countries were found to have strengthened their capacity to align their programs with the national development strategies of developing countries and by doing so were able to support capacity development (OECD 2012). While a single result might be disappointing in terms of overall expectations, it is also important to see its achievement in light of the need to shape the 'enabling environment' which is critical to subsequent success.

4.2 Research Process

According to Alston and Bowles (1998), a research process is always employed by researchers aimed at answering a research question. Although the research process may be less straightforward in practice; it is often presented in linear formats. In presenting research findings, usually in the form of a research report or published journal article(s), the researcher(s) is obligated to explicitly detail the research methodology used.

Identification of Research Problems

A problem, or problems, must be able to be researched, must have the ability to be duplicated in practice and must, in addition, be supported by existing literature. For the sake of developing a base of existing evidence, the problem(s) that need to be addressed may either be local, national or international (Alston, 1998).

Search for the Existing Literature Base

Before conducting the actual research, the problem should be supported with an extensive search of the useful and relevant literature using the internet, data bases, and texts as well as expert sources. The literature should not only be broad, but also in depth so as to provide peers with proof that a comprehensive search of the existing research problem area (Alston, 1998) has been performed so can withstand any challenges to its veracity.

Literature and Critical Appraisal

A review of the literature should be employed in a systematic manner so as to create and develop a critical appraisal framework (Alston, 1998).

Developing Research Questions and/or Hypothesis to be Tested

It is important to develop a specific research question and/or hypothesis from the literature in order to provide a specific direction for the research, with the aim of providing answers to the question or hypothesis posed (Alston, 1998).

Theoretical Base

A theoretical base may be employed by the researcher in order to examine the problem, and is usually seen in higher levels of research. Theoretical base in health as well as social care fields might be derived from social sciences, anthropology or psychology (Alston, 1998).

Sampling strategies

Sampling is the method used in selecting people, objects or events for study in research. Probabilities as well as non-probability sampling strategies help the researcher target data collection techniques. There is always a need for a specific sample size, which is often determined through calculation, but can also be obtained through composition (Alston, 1998).

Data Collection Techniques

These include the approaches as well as the tools used to collect data for the purpose of answering the research question or the hypothesis. In some cases, two or more techniques may be used. The most common techniques are interviews and questionnaires (Alston, 1998).

Approaches to Quantitative and Qualitative Data Analysis

The approach involved in data analysis may either be quantitative or qualitative and is usually dependent on the kind of data collected (Alston, 1998).

Interpretation of the Results

The results are to be interpreted accordingly as they not only help draw the conclusions, but they also answer the research question or hypothesis. In addition, implications for practice, as well as any further research to be done are decided on acknowledging the limitations of the research (Alston, 1998).

Dissemination of Research

The research, as well as the results, can then be presented in the form of electronic and print forms through articles, written reports, conferences and papers (Alston, 1998).

4.3 Analysis of data

Jeffrey (2013) identified six archetypical analyses, which can be used in the data analysis, which include descriptive exploratory analysis, inferential analysis, predictive analysis, causal analysis, and mechanistic analysis.

Descriptive Analysis: This involves a quantitative description of the key features of a given collection of data. It is usually the first type of data analysis carried out on a data set and is often applied when dealing with large volumes of data, such as census data sets. In this analysis, the processes of description and interpretation are different steps. The two types of statistical descriptive analysis are the univariate and the bivariate.

Exploratory Analysis: This type of analysis investigates the previously unknown relationships in a data set. Exploratory models are important for the discovery of new connections as well as for defining future questions and studies. Exploratory analysis does not usually provide definitive answers to questions at hand; neither can it be used for predicting and/or generalizing. It can be applied to census data sets as a Convenience Sample Data Set, which is typically non-uniform and is a random sample in which many variables are measured.

Inferential Analysis: It employs a sample of data, which is relatively small to say something concerning a bigger population. Inference is, generally, the goal of most statistical models, and involves the estimation of both quantity of interest and the uncertainty about the estimate. The inference is heavily dependent on the population as well as on the sampling scheme. The type of data sets that inferential analysis is applied to include a Cross Sectional Time Study, Retrospective, and Observational Data Sets.

Predictive Analyses: These include a number of different methods that use current as well as historical facts in order to make predictions about future events. The models basically use the data on some given objects in order to make predictions of values associated with another object. The accuracy of the prediction heavily depends on measuring the correct variables. Although both better and worse prediction models exist, it is always advisable to use more data as well as a simpler model. These models can be applied to Prediction Study Data Set.

Causal Analysis: This helps in finding out how changes in one variable affect another variable. It is usually implemented in randomized studies and is applied to Randomized Trial Data Sets.

Mechanistic Analysis: This helps in understanding the specific changes in variables which result in changing other variables for the individual objects. It is usually extremely hard to make inferences in mechanistic analysis, unless a researcher is dealing with simple situations. This analysis is often modeled using a deterministic set of equations (engineering/physical science), and if the equations are known and the parameters are unknown, the unknown parameters may then be inferred with data analysis. Mechanistic analysis is also applied to Randomized Trial Data Sets.

4.4 Analyzing Information

According to Shapiro (2001), a researcher may find that there is a large amount of information during monitoring and evaluation (M&E), so must find a way to make sense out of it or find how to analyze the information. Shapiro (2001) further explains that if the researcher is using the services of an external evaluation team; it is always the case that the analysis is carried out by that team. Although sometimes in evaluation, but certainly in monitoring, the researcher may be forced to carry out the analysis in person.

Shapiro (2001) defines analysis as "the process of turning the detailed information into an understanding of patterns, trends, interpretations." In a project, or an organizational context, the initial point for analysis is usually very unscientific and it is the researcher's intuitive understanding of the main theme(s), which come out in the processes of information gathering (Shapiro, 2001). Once the main themes have been identified, the researcher can work through the information with ease by structuring and organizing it. The researcher should follow this up by writing up the analysis of the findings as a basis for making conclusions as well as recommendations. The process should follow the following criteria:

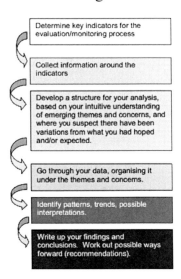

Figure 3.2: Criteria for analyzing M&E (Shapiro, 2001)

4.5 Development Assistance Committee Criteria

The OECD's Development Assistance Committee (DAC) put a number of principles or evaluation in place in 1991, to guide its member states. These principles have since been used in development evaluation by separating them out to form five different criteria which include: relevance, efficiency, effectiveness, impact and sustainability.

Owing to the need to evaluate complex emergencies, the five criteria have since been adapted and then expanded into seven criteria, which include relevance, connectedness, coherence, coverage, efficiency, effectiveness, and impact (OECD-DAC, 2000), which are supposed to be used hand in hand in a complementary fashion.

Relevance: This shows how well the aid activities are suited to the policies and the priorities of the donor, the recipients and the target group.

Effectiveness: This measures how well the aid activities attain their stated objectives.

Efficiency: This measures the outputs, both qualitative and quantitative, with respect to the inputs. In an economic context, the term efficiency indicates that the aid uses the least costly resources possible to attain the desired results. This calls for a cost comparison between alternative approaches in order to attain the same output, so as to adopt the most efficient process.

Impact: This is intended to cover/expose both positive and negative changes that have resulted from the development intervention, intended or unintended, directly or indirectly.

Sustainability: This measures the possibilities for the continuity of the benefits of the aid activity, even after the funding by the donor has been withdrawn or terminated. The project(s) should not only be financially sustainable, but also environmentally sustainable. 'Sustainability', in development, continues to be controversial, inconsistent and imprecise depending on who is using the term. No standard and universally acceptable definition currently exists. Consensus remains elusive and polarizing"

In most cases, there is always some overlap among these criteria and the same set of data can be used in different criteria. The Active

Learning Network for Accountability and Performance (ALNAP) in Humanitarian Evaluation and Learning Portal (HELP) identified eight cross-cutting and important themes, which should always be carefully considered by the evaluators when using the DAC criteria: human resources, local context, the participation of primary stakeholders, protection, coping strategies and resilience, HIV/AIDS, gender equality, and the environment. It is not necessary for an evaluation to include all of the eight themes, but a rationale should be taken into consideration if any of them is to be excluded (OECD-DAC, 1998).

Although the DAC criteria are widely used in evaluation, they are often used mechanistically. They are a very important guide when framing questions as well as for designing evaluations, but a researcher has the freedom to be creative in the process of evaluation even when relying on them (OECD-DAC, 1992).

5. Methodology

The approach used in developing this manual as far as possible has focused on:

- Presenting a contextual meaning of the subject matter based on the author's experience and available literature review;
- Developing an example to illustrate the applicability of the subject matter and
- Presenting an appropriate and relevant case-study when available.

All these have been presented, as far as possible, in non-technical language. Every effort has been made to stay focused on the subject and avoid unnecessary deviations from the relevant substance.

6. Target Audience

As has already been mentioned elsewhere, the author has tried to gear the subject matter to interest as many readers as possible. Groups that will find this document useful include staff of academic institutions, government administrators, Non-Governmental Organizations (NGOs) and Program Implementing Partners, such as the staff of organizations nominated as Implementing Partners for Multi and Bilateral funded programs.

7. Results-based Management (RBM)

It seems appropriate to start the technical part of this manual with a brief look at Results-based Management (RBM) -- its scope, applicability and validity. RBM is quite broad, inclusive and practical as demonstrated by the many experts and different stakeholders.

The concept of RBM is not new. Indeed, it underlies all the effort that humans consciously undertake in order to achieve desired results. For example, countries go to war to win (outcome); parents bring up their children well with the goal of their becoming law-abiding citizens; children go to school (elementary, middle school, high school, community college, university) to successfully graduate (outcome); farmers plant crops in anticipation of the resulting good harvests; people go to the gym to lose excessive weight (outcome); businesses are established in order to make profits; and the list continues. All these interventions have the common objective to transform the inputs into intended and sometimes unintended results (outputs, outcomes, and goals) or strategic objectives (SO).

The universality of this concept makes it a useful basis by which to streamline one's effort in order to optimize proposed interventions. From a program development perspective, definitions have generally limited RBM to the key results presented in a relevant program design framework (PDF). This approach, while correct, seems to limit the importance and the significance of the scope of RBM. Various organizations and groups have all developed different definitions. For example, OECD/DAC defines RBM as: "A management strategy focusing on performance and achievement of outputs, outcomes and impact".

Here, the definition of RBM has intentionally been expanded. It has also been deliberately presented from a generic perspective. Such a strategy creates room for every stakeholder and facilitates the "portability" of lessons learned. It also enhances the versatility of this concept.

In this primer, I have attempted to define a more holistic meaning of RBM as follows: RBM can be considered to be a hierarchical framework of mutually complementary components (program design framework, monitoring and evaluation (M and E), and data management (DM)) with synergistic dynamics that collectively yield the intended and sometimes unintended results. The beauty of RBM is that there is no absolute "goal standard". The unavailability of an acceptable "turn-key" system is confirmation of the dicey nature of RBM. Each intervention is different, and as such, RBM serves as a unique framework with the common potential of promoting efficiency (inputs, outputs) and accountability (effectiveness). Effective and evidence-based performance monitoring contributes substantially to the successful achievement of RBM. It is my hope that such a concept will not only serve a wide audience on the one hand, but also facilitate program design and implementation on the other.

While the target audience of this manual remains quite broad – bilateral and multilateral agencies, academic institutions and program implementing partners (PIM), it is my hope that an RBM document can also be used by anyone who is interested in an improved understanding of the concept. To this end, as far as possible, I have tried to minimize utilizing technical jargon where appropriate. Such an attempt will hopefully expand the target audience base and generate more interest, especially, among the members of the program development communities. Finally, the current requirement by donors for accountability and results has provided compelling evidence for the need for such a booklet at this time.

8. Monitoring and Evaluation (M&E)

8.1 Basic Principles of Monitoring and Evaluation

The term 'monitoring and evaluation' is sometimes misconstrued, as if they were a single thing. Yet monitoring and evaluation are, in reality, two different sets of organizational activities, which while related are quite dissimilar.

Monitoring of a project/program is the methodical collection and analysis of information pertaining or derived from observations of the work processes during the progress of a project.

The main aim of monitoring is to diagnose whether the work (work management plan) is working out as expected with the resources allocated, and/or to use the data gathered to improve, if necessary, the efficiency of those processes, as well as to perhaps positively affect the effectiveness of a work output/outcome project. Monitoring is often based on measuring achievement against set targets, as well as the successful completion activities that are planned and budgeted for during the planning phases of work. It also helps keep the progress of work on track/schedule, or by proposing/making corrections to the work plan's budget or schedule to get it back on track by making management aware whenever things are going wrong.

If monitoring is done properly; it lays down a useful foundation for the evaluation that will follow. Monitoring also enables a researcher to determine if the resources available (inputs) are sufficient and whether they are being optimally utilized, if the project/program staff's capacity is appropriate and sufficient, as well as if the realization of both the project/program outputs and outcomes are progressing as planned.

Monitoring is an internal function in any organization or project and involves the establishing of indicators of efficiency, effectiveness and impact. Monitoring includes the setting up of systems used in the collection of information related to the selected indicators, as well as collecting and recording the data/information, analyzing the data/information, deriving conclusions, formulating recommendations, and, lastly, in applying those recommendations selected and resourced to day-to-day operations and overall program management.

Evaluation of a research project involves the comparison between actual project/program impact and it's agreed on strategic plans. Evaluation is also concerned with what is set out to be done, what has to be accomplished, and how it is to be accomplished. Evaluation can be formative, meaning that it can take place during the progress of the project/program with the main intention of improving/altering the strategy thrust of the project. It can also be summative, that is, it can draw lessons from a completed project/program, which may be used in subsequent program planning exercises.

The common aspect between monitoring and evaluation is that they are both geared to learning from what is being done and the manner in which it is being done, by focusing on efficiency, effectiveness and impact. Efficiency helps to determine whether the input is appropriate when compared to output. The input can usually be valued in terms of time, money, staff/consultants deployed, and equipment and supplies allocated among others. The effectiveness of a research project measures the extent to which the specific objectives are achieved. Measuring the impact of a development program or project helps determine whether the factors such as policies and resources deployed made a difference to the problem being addressed; that is, was the strategy employed useful/ vital to achieving the intended outcomes and the overall goal.

The following are the most commonly used terms under M&E:

i. Inputs: include the human, financial, technical and material resources used in developing the intervention, and can be categorized as management structure, technical expertise, equipment and funds.

ii. Activities: include actions taken or work plans developed and work performed, such as training workshops conducted, coordination meetings organized, procurements made, quality of physical work undertaken, and monitoring conducted.

iii. Outputs: include the capital goods, products, and services resulting from development intervention, such as the number of people trained, the number of workshops conducted, and the number of widgets constructed / manufactured.

iv. Outcomes: changes observed/achieved pertaining to short-term as well as medium-term effects traceable to an project/ program intervention's outputs, such as on productivity due to increased skills derived through training; additional demand for existing or new services experienced; sales or distribution of desired program products to intended beneficiaries, and new employment opportunities created; .

v. Impact: the program's long-term consequences, which may either have positive and/or negative effects such as an improved standard of living for a target population.

8.2 Monitoring and Evaluation Plan

The Monitoring and Evaluation plan (commonly abbreviated as the M&E plan) is a document, which is employed by the project team that can help plan and manage all Monitoring and Evaluation activities in the whole cycle of a particular project. It is necessary for it to be shared and utilized among all the stakeholders, in addition to being shared with program donors if so desired.

A M&E plan enables those responsible for program management to keep track of what indicators should be monitored, when to take measurements, where to monitor and when to conduct evaluations. The M&E plan, which may take the form of a typical MBO-work plan that is specific to activities based on Monitoring and Evaluation, generally includes the following:

i. Goals and objectives of the program and matrix of outputs and outcomes associated with these.
ii. Questions to be asked and methodologies to be employed in conducting and monitoring M&E.
iii. iii Work-plan or Implementation plan.
iv. Iv Matrix of monitoring indicators as well as their intended results
v. Proposed timetable for all the activities related to monitoring and a separate indication of when evaluation(s) [baseline survey; mid-term or mid-cycle; final] will take place.
vi. Samples of pretested monitoring and evaluation instruments to be used in gathering data.

A good M&E plan should be rigid on one hand and flexible on the other. It should be rigid in the sense that it should be well planned and thought out, and flexible in the sense that it can account for any change, which could improve the monitoring and evaluation practice. Both rigidity and flexibility are equally important because of the ever-changing program environment and because of certain fast moving conflict environments.

The M&E plan enables all stakeholders involved to have a common frame of reference that details all the M&E activities that will occur as the project progresses. The M&E plan also helps signal and identify changes in data flows as the project/program progresses over a given period and helps funnel data into flows that contribute to the subsequent evaluation processes.

To sum up, an adequate M&E plan is very useful in management and its development is a 'best practice'. It should be employed as a reference tool throughout and should frequently be updated so as to incorporate unplanned interruptions or shortfalls of resources such as materials, manpower and money and illustrate how such unplanned events might affect the production of outputs, the quality and timeliness of desired outcomes. The status of constant variables inherent to the management/operational work-plan should be updated rigorously by specific staff and according to the schedule in the management plan. These updated Monitoring indices should be presented in frequent biweekly or monthly staff meetings. The relevant audience should include the program management and the exact personnel specific for each activity or intervention. This group would then follow the approved methods and use preauthorized sources of data collected

USAID requires that the program's official monitoring and evaluation (M&E) plan should be completed one month after the signing of a program or project contract. As the M&E plan is intended to be used in an organized set of coordination meetings, its creators should be those involved in the program and should also include the strategic partners. This ensures that those who designed the M&E plan are also the users. The other advantage of this participatory approach is that it ensures the project team has the support of the major stakeholders and it also serves as a safeguard against arbitrary intervention by strategic partners and/or other stakeholders.

There is no particular standardized structure or template for an M&E plan, but an effective M&E should include the parts listed below:

1. Executive Summary
2. Project Background
3. M&E Planning
4. Monitoring and Evaluation Systems
5. Project Risk Matrix
6. M&E Information Map
7. M&E Work Plan Matrix
8. M&E Timetable
9. References
10. Annexes

The above outline can be modified with the consensus of the stakeholders to best fit the communication, program, and organizational needs of these stakeholders. Other documents may also be used as reference aids when writing the M&E plan. These include:

i. Project/Program Proposal
ii. Logical Framework
iii. Results Framework
iv. Checklist for M&E Plan

These supporting documents usually have information that is included in the M&E plan. They also make referencing much easier as well as ensure continuity between all project-related documents. The depth/detail of information included in the M&E plan should be more specific than that found in the supporting documents and some cases may, occasionally, call for some (contextual) revision to the goals and objectives as stated in the original documents.

It is important to consider carefully both resource and budget constraints when making an M&E plan. One of the greatest of these is the cost of data collection, which is usually divided into human resource expenditure and other financial costs such as computer time, printing, copying, etc. A program manager should be able to tell if the time proposed as well as the resources made available would be sufficient to cover all the described/planned M&E activities. A program manager should also model whether the value derived from the data collection process is worth the budgeted expenses of the data collection process.

A properly executed M&E plan is usually very helpful in the identification of problems, in the planning and the implementation, in reviewing the progress of the project in terms of intended outcomes, as well as in making any necessary adjustments within the project.

Most organizations consider M&E as more of an internal process requirement rather than as a strategic management tool. Although, it is the duty of management to provide stakeholder/donor reassurance that their money is being well spent, the primary purpose of a M&E plan should be to help the organization (or the project) monitor how it is progressing against the objectives, to show if it is having the intended impact, whether it is working efficiently, and if there is a need to improve the processes and then show ways and means of improving it. All these reasons make the M&E tools the key pillars of a strategic framework which includes the vision, the problem analysis and the project/program's strategic value.

8.3 Program Indicator Screening Matrix (PRISM)

8.3.1 Introduction

Historically, development program implementation has been plagued with a complex set of challenges, which vary from the program design stage to implementation plans and processes, to long-term sustainability. A key element of program design and implementation has been the selection of the relevant indicators and the capacity to optimize its robustness and mitigate the prevalence of bias. These challenges continue to influence effective program management initiatives. Attempts to address some of these issues vary from program to program. For example, what some program designers may identify as low-level indicators in practice, sometimes represent higher-level indicators. Such a scenario could misrepresent or affect a program's potential results.

The PRISM tool is a table/matrix which analyzes each indicator. This effort is executed by a team of experts who are selected and grouped based on their relevant expertise. An initial attempt is made to clearly describe the matrix, its limitations, and how it assists in addressing some of the challenges faced by program implementing partners in establishing meaningful indicators. The final outcome of this exercise is a consensus or high degree of concordance (as opposed to discordance) among the program leadership and staff (team) members.

The presentation outline of the PRISM is as shown below:

    i. Introduction
   ii. Objectives
  iii. Relevance
  iv. Target Audience
   v. Evaluation Life Cycle (ELC)
  vi. Program Design Framework (PDF)
 vii. Themes
viii. PRISM: A Composite Score Framework
  ix. Lessons Learned

The PRISM should highlight the following points:
   i. Evaluation – Demand driven
   ii. Participatory
   iii. Inclusive
   iv. Bottom Up Strategy (BUS)
   v. Consensus - Based
   vi. Random Thematic Subgroups
   vii. Intra-Thematic-Group Concordance
   viii. Inter-Thematic-Group Concordance
   ix. Bar (Gold Standard vs. Effective)
   x. Binary Outcome
   xi. Delphi Methodology
   xii. Mapping
   xiii. Scope: Africa, Asia and Pacific Island Countries

8.3.2 *General Objectives*
   i. To strengthen the knowledge of IPs, PMs and other key stakeholders and emphasize sustainable engagement in program management and implementation processes.
   ii. This is in an attempt to address existing nuances, highlight the synergies that exist among the different result levels of the SFW and hence facilitate a common ground between potential evaluators and different interested parties.

8.3.3 *Specific Objectives*
   i. Streamline the monitoring plan by improving indicator causal links at all result levels;
   ii. Mitigate duplication of indicators;
   iii. Establish authentic contributions between different result levels;
   iv. Establish meaningful synergies among different result levels: no lower level result can contribute to more than one upper-level result;
   v. Strengthen the program design;
   vi. Promote a common understanding among key actors; and
   vii. Minimize cost and optimize the number of indicators included in the program.

8.3.4 *Relevance of the PRISM*
  i. Improve intended and unintended intervention results and make foreign aid more focused with evidence-based results;
  ii. Establish more effective, continuous and sustainable synergies among frontline forces, IPs, Funding Agencies, Stakeholders and Beneficiaries.

8.3.5 *Audience Targeted by the PRISM*
  i. Funding Agencies;
  ii. IPs;
  iii. Program Managers;
  iv. Relevant Stakeholders;
  v. Evaluators;
  i. Development Partners.

8.3.6 *Evaluation of the Life Cycle*

i. Demand Recognition
The evaluation process demands that its attributes be identified and preserved by all the project's stakeholders.

ii. Evaluation Team Identified
It is important to identify the best team to carry out the evaluation process. The team may either be drawn from the staff of the organization (self-evaluation) or have an external team carry out the evaluation process (external evaluation). In some instances, the project staff and the intended beneficiaries may be included in the team (participatory evaluation).

iii. Inception Report Developed
An inception report that summarizes the review of the document undertaken by the evaluation team should contain about 20 to 25 pages. This report is used to set out the project's objectives, the suitable approach to be used, the methodology, the working program as well as the study team for the present study.

iv. Evaluation Process Implemented
The inception report should be revised where necessary and the implementation of the evaluation process should be carried out.

v. Draft Report Developed and Presented

It is necessary to develop a draft evaluation report before submitting the final evaluation report. This report should be presented to the stakeholders, who should give their feedback after the presentation for further action (revision).

vi. Final Report Developed and Submitted.

The final evaluation report should be developed and submitted after making the necessary revision to the draft evaluation report based on the feedback of the stakeholders.

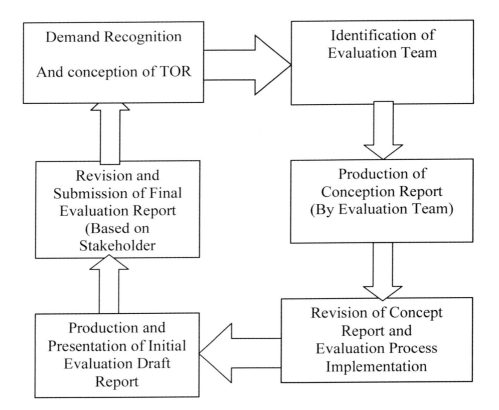

Figure 4: Evaluation Life Cycle (Lainjo, 2013)

Shapiro (2001) illustrated the effect of M&E using the cycle shown in figure 5. It is worth noting that monitoring as well as adjustments should be carried out several times before one is ready to evaluate and replan.

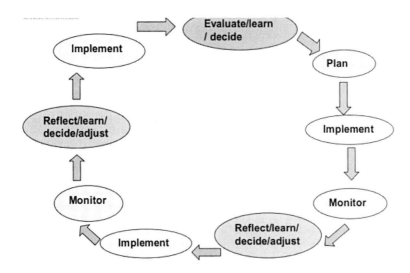

Figure 5: M&E Data Management Cycle (Shapiro, 2001)

It is important to use M&E systematically because there is no guarantee that it will solve the problems without the relevant and interested parties undertaking the work necessary to formulate this evaluative process. Monitoring and evaluation can help identify the problems as well as their causes, suggest possible solutions to the problems, raise questions concerning the strategy and assumptions, provide and encourage action on information and insight, as well as increase the possibility of being able to make positive development difference. Various methods of carrying out evaluation exist. Some of the most common methods include:

Self-evaluation: An organization or a project management team assesses the program through the administration of structured questionnaires, or through self criticism about how well it is doing -- as a means of learning as well as improving practice. This calls for the staff of an organization or team to be very honest and self-reflective in order to effectively achieve this.

Participatory Evaluation: This requires as many members of the project/program team as possible to participate. It is more of an internal evaluation, whereby project staff and sometimes the intended beneficiaries are called upon to work together on the evaluation.

Outsiders may only be invited to act as process facilitators, but not as evaluators.

Rapid Participatory Appraisal: This is a qualitative method of evaluating, which involves a review of secondary data, semi-structured interviews, testimonies of key informants, direct observations by the evaluators, group interviews, diagrams, games, maps and calendars. It was originally used in rural areas, but has also extensively been applied in other communities as well.

External Evaluation: This involves the choosing of an outsider or an outsider team to carry out the evaluation.

Interactive Evaluation: This involves free flowing participation between the evaluation team, or outside evaluator, and the organization or the project that is being evaluated. An insider may sometimes be included as part of the evaluation team.

8.3.7 *Program Design Frameworks*

Type	Result Levels	Agencies
Logic Framework (Logframe)	☐ Impact ☐ Output ☐ Outcome	UN, CIDA, EU, AusAID, DfID, WB
Strategic Objective	☐ Strategic Objective ☐ Program Objectives ☐ Program Sub-objectives	USAID

Table 1: Program Design Frameworks (Lainjo, 2013)

8.3.8 *Logframe*

Below is a diagram showing IF-THEN chain results:

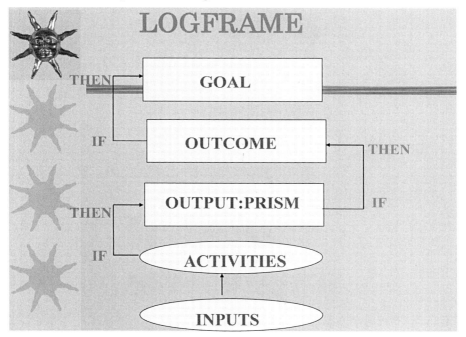

Figure 6: Logframe (Lainjo, 2013)

Testing Internal and External Logic: *If* the OUTPUTS are delivered through planned ACTIVITIES using the relevant INPUTS and the corresponding ASSUMPTIONS at the OUTPUT, OUTCOME and IMPACT levels; they remain valid. So that the desired OUTCOME will materialize leading to the intended IMPACT.

The goal of the logframe is described in terms of quality of life.

The outcome of the logframe is determined by asking the question "How will this goal be achieved?" and is described in terms of use, attitudinal change, and political commitment. For example, access to comprehensive RH services increased. The other example is the utilization of age/sex disaggregated data improved.

The Output is concerned with 'deliverables', time bound strategies. For example:

i. Increased availability of comprehensive RH services
ii. Improved quality of RH services
iii. Improved environment for addressing practices that are harmful to women's health
iv. National development plan and sectoral plans in line with ICPD/POA
v. Increased availability of sex disaggregated population related data
vi. Increased information on gender issues

The Inputs include resources such as staff, overhead costs, materials, equipment and other project costs such as data processing, report writing, printing and dissemination of results.

Component	Description	Monitoring	Evaluation	Criteria*
Goal	Described in terms of quality of life	No	Yes	Sustainability Impact
Outcome	Described in terms of use, attitudinal change, and political commitment	No	Yes	Effectiveness Relevance
Output	Concerned with "deliverables", time bound strategies	Yes	Yes	Efficiency
Activities		Yes	Yes	Efficiency
Input	Include resources such as staff, overhead	Yes	Yes	Efficiency

	costs, materials, equipment			

Note: * = OECD/DAC Working Group Criteria

Table 2: Strategic Framework and M & E Summary Table

Criteria	Description
Sustainability	Measures the possibilities for the continuity of the benefits of aid activity, even after the funding by the donor has been withdrawn or terminated
Impact	Intended to cover/expose both positive and negative changes resulting from the development intervention
Effectiveness	Measures how well the aid activities attain their stated objectives
Relevance	Shows how well the aid activities are suited to the policies and the priorities of the donor, the recipients and the target group
Efficiency	Measures the outputs, both qualitative and quantitative, with respect to the inputs

Table 3: OECD/DAC Working Group Criteria Definition

8.3.8.1 *M & E – Analysis Case Study*

Background

In a typical evaluation term of reference, evaluators are often required to assess the extent to which inputs (human/financial resources, materials etc) contribute to the output during and after program implementation.

The case study (human resources) represents partial actual findings of an evaluation exercise conducted in a funding agency's office. In the office, there was also a consensus by many staff members that, with regard to work loads, while there was an increase in numbers of staff as a result of the restructuring exercise, the increase had not been proportionate to the work loads.[1] Staff members continue to be faced with additional tasks that have cumulatively resulted in additional and higher stress levels. The reluctance of staff members to assist their colleagues in accomplishing certain tasks during high demand periods has only exacerbated an already dire situation.[2] In Figure 1, an attempt has been made to establish a possible link or association between "absenteeism" and work performed from 2008 to 2010. In this context, the number of sick days will be considered a proxy for stress levels. This analysis will help explain some of the concerns raised by staff and help management introduce an informed decision strategy that will contribute to improving staff performance.

Analysis

Based on the average number of sick days taken by staff during the period 2008 to 2010, there is evidence, as illustrated in Figure 1, that there was an increasing trend during that interval.[3] The figure also shows that this trend peaked in 2010. While one might suspect that the increase in sick days might have been caused by other reasons, there is the likelihood that the significant increase of "absenteeism" in 2010 could have also been caused by higher stress levels possibly related to higher staff workloads as indicated earlier. The figures also suggest that during the last year, the mean number of absent days was also close to ten, which is the maximum payable number of sick leave days granted to staff. The figure also indicates that employees took over twice as many more sick days in 2010 than in 2008. If these observations remain valid, then staff concerns about heavier workload are consistent with the distribution of the number of sick days taken.[4]

---

[1] This was an exercise in which the Agency decentralized its activities. The process was applied to regions all over the world.
[2] The disenchantment among staff continued to compromise productivity. This was confirmed by some members during discussions with the team.
[3] Funding Agency Statistics.
[4] Adjustments were made for two outliers: one in 2009 (115 days) and one in 2010 (159 days). In each case, the average figure excluding these cases was calculated and assigned to these outliers.

Figure 7: Mean distribution of staff sick leave from 2008 to 2010

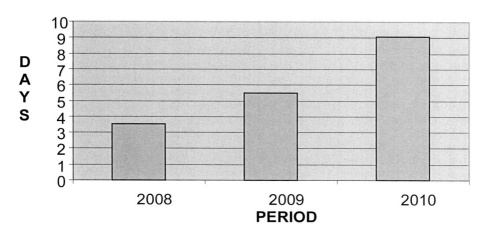

The median has also consistently confirmed the trend by demonstrating that:
- ✓ Half the members of staff took four (median) or more days of sick leave in 2008.
- ✓ Half of Agency's staff took at least five (median) days of sick leave in 2009 and
- ✓ Half of the employees took eight (median) or more days of sick leave in 2010.

Conclusion

There is sufficient evidence here to suggest that staff stress levels (based on sick leave days) progressively increased during a three year period. Further to this, there is an obvious inverse relationship between level of effort and sick leave vacations. Hence, productivity was compromised in this circumstance; thereby, negatively affecting efficiency at the input level. Therefore, the likelihood exists to believe that this trend cumulatively and negatively affected staff performance. Senior management is thus requested to intervene with a more informed strategy in order to assist staff improve their productivity. This strategy also needs to effectively be monitored to make sure that the appropriate results are achieved.

The table that is used to establish the number of acceptable indicators is made up of as many ROWS as there are indicators and TEN COLUMNS. The first row represents descriptions of each column. For example, in row one, column one, the relevant thematic area, result-level and indicator are filled in. In the next eight columns (still on row one), the respective criterion is filled in that will be used in screening the indicators. In the row below and subsequently, a table of binary elements is shown, i.e. zeros and ones (0, 1). The former represents a corresponding indicator, which does not satisfy the criteria whereas, the latter, is a corresponding indicator which fulfills the criteria. The same process applies to the all the criterion and corresponding indicators. Column seven summarizes the scores in terms of number of yeses (or 1). The seventh column is the final score attained by each indicator, which is represented as a percentage of positive responses in the row. The last column is the final outcome, which tells us if, based on the scores (1s), one should go ahead and recommend the indicator or not. The 'gold standard' for this exercise is 100%. That would be an indicator that scores positive responses in all the criterion so it qualifies for implementation automatically.

In short, the matrix should contain the following:

i. An R by C Matrix where
   i. R = Number of Thematic Indicators and
   ii. C = Six Screening Criteria.
ii. Each Indicator is cross-tabulated with each criterion;
iii. The intersecting cell is filled with either a "1" or a "0";
iv. The former, if the indicator satisfies the criterion; and the latter if it doesn't;
v. Exercise continues until all indicators are screened;
vi. A corresponding final score (%) per indicator is established for each row. These are used to establish Group Concordance;
vii. Thematic Group and Sub Groups agree on effective %;
viii. Each Subgroup is made up of Moderator, Rapporteur and Team.

Program Indicator Screening Matrix (PRISM)

Thematic Area: Results Level: Goal, Outcome, Output indicators	1 Specificity	2 Reliability	3 Sensitivity	4 Simplicity	5 Utility	6 Affordability	7 Total Yes	% Score	Implemented Yes/No

1. **Specificity** - Does it measure the result and contribute to ONLY 1 higher level indicator?
2. **Reliability** - Is it a consistent measure over time?
3. **Sensitivity** - When the result changes will it be sensitive to those changes?
4. **Simplicity** - Will it be easy to collect and analyze the Data?
5. **Utility** - Will the information be useful for decision-making and learning?
6. **Affordability** - Can the program/project afford to collect the Data?

Table 4: Program Indicator Screening Matrix (PRISM) (Lainjo, 2013)

8.3.10 *Themes of the PRISM*

    i. Health
    ii. Education
    iii. Environment
    iv. Governance
    v. Poverty
    vi. Judiciary
    vii. Agriculture
    viii. Social Security and Protection

8.3.11 *Criteria of the PRISM*

As this is a composite analysis, we need to remember that a final outcome is only valid when all these criteria are considered simultaneously (Appendices 2b to 2e). That is, the outcome identified in the last column. What happens if no indicator satisfies all these conditions? The answer is simple. Before all the subgroups begin their assignment, there must be a consensus established by the team with regard to an acceptable level. For example, the team could agree before the exercise starts that any indicator that scores 70% (total positive responses divided by sum of positive responses and negative responses) or another decision level, will be considered acceptable. Sometimes, this bar can vary. For example, if the team recognizes that a certain threshold tends to admit too many redundant indicators; the bar can be raised higher in order to further refine the choices.

The following paragraphs attempt to define the meaning of each criterion as it applies to the matrix.

Specificity: This refers to the likelihood of the indicator measuring the relevant result. In other words, is there a possibility that the result the indicator represents does not represent exactly what we are looking for?

Reliability: This criterion is synonymous to replication. That is, does the indicator consistently produce the same result when measured over a certain period of time? For example, if two or more people calculated this indicator independently, would they come up with the same result? If the answer is 'yea', then the indicator has satisfied that condition and, hence, a 'one' is entered in that cell or else zero entered.

Sensitivity: A test to assess the stability of an indicator. For example, does the indicator continue to deliver the same result with a small variation of either the numerator or denominator? How does the result change when assumptions are modified? Does the indicator actually contribute to the next higher level? For example, an indicator at the output level accounting for one at the outcome level will yield a misleading result. If the same indicator accounts for two or more result levels simultaneously; it is not stable. As indicated earlier, any indicator that satisfies a criterion is given a one in the corresponding cell or else a zero.

Simplicity: A convoluted indicator represents challenges at many levels. Hence here, an indicator that is easy to collect, analyze and disseminate is preferred. Any indicator that satisfies these conditions automatically qualifies for inclusion. The zero/one process is then followed as indicated above.

Utility: This refers to the degree to which information generated by this indicator will be used. The objective of this criterion is to assist in streamlining an indicator in an attempt to help the decision-making make an informed decision. This can be stipulated either during the planning process or during the re-alignment process. The latter represents occasions when an organization is evaluating the current status of its mandate.

Affordability: This is simply a cost-effective perspective of the indicator in question. Can the program/project afford to collect and report on the indicator? In general, it takes at least two comparable indicators to establish a more efficient and cost-effective one. The one that qualifies is included at that criterion level. Then, the same process as outlined above is followed.

Inclusion: The penultimate column (8) simply represents the composite score. The total number of positive responses is divided by the total number of criterion (in this case, seven) and multiplied by 100 to produce the relevant score for each indicator. During this process, each indicator is then classified as either accepted (if it scores 70% or more in this example) or is otherwise rejected.

8.3.12 *Implementation of the PRISM*

     i. Theme Identification
     ii. Thematic Group Selection
     iii. Random Thematic Sub Group selection
     iv. Selection of Sub group Moderator and Rapporteur
     v. Individual Thematic Subgroup member Scoring
     vi. Establish Intra-Thematic-Subgroup Concordance
     vii. Establish Inter-Thematic Group Concordance
     viii. Conduct Thematic group Plenary
     ix. Establish Group Consensus
     x. Select Final Set of Indicators

8.3.13 Algorithm of the PRISM

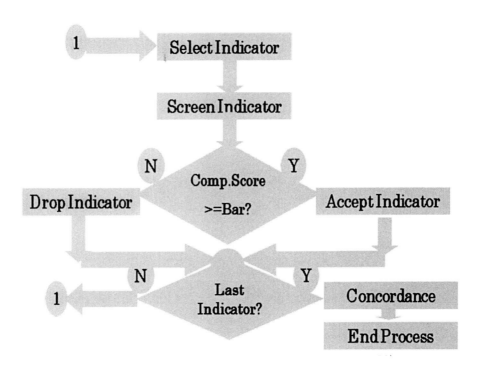

Figure 8: Algorithm of the PRISM (Lainjo, 2013)

Descriptive Procedure of PRISM Algorithm

10 For Any Theme;

15 Are there any more themes? NO →→ Step 70

    20 Select thematic Indicator;

    25 Has the last thematic Indicator been Screened?

YES→→ Step 70

30 Screen thematic Indicator against Criterion; {Process continues through every Criterion}

35 Has Indicator been Screened against ALL criteria?

YES>→→ Step 45

40 →→ Step 30;

45 Compute Indicator Composite Score {(Sum of "1"/(Sum of ("1" + "0"))};

50 Compare Composite Score with pre-defined BAR;

55 IF Composite Score >= BAR, ACCEPT Indicator >>>>>>Step 20;

60 IF Not, DROP Indicator →→ Step 20;

70 Calculate Group Concordance;

80 END PROCESS

The following lessons can be learned from the PRISM:
    i. The importance of team composition homogeneity
    ii. Consensus-building required
    iii. No more than ten members per thematic subgroup

iv. Solid knowledge of theme essential
v. The importance of time management
vi. Framework useful for pre-program implementation
vii. Also essential during Mid-Term-Review (MTR)
viii. Active involvement of top management critical
ix. Feedback provided to all active teams required
x. Useful initial contact tool for evaluation team and relevant program key players.

9. Application of Data Management Framework:

One of the most important aspects of a good data management framework is that it enables the researcher, as well as institutions, to meet their obligations to their funders and to make data available for sharing, for validation and for reuse. To achieve this, the management of data derived from research should be carried out transparently from the onset by employing various stages, which include planning, collection of data, analysis of data, publication of the study results, archiving the data and enabling its later recall and reuse.

A good data management system requires the contribution of all those party to the exercise, which implies each player has a specific job description. The roles played by institutions include establishing as well as promulgating policies and procedures, the provision of the necessary and sufficient infrastructure as well as service support. Primary researchers have the responsibility of managing their data within the institutional framework in line within the requirements not only of the Funding Agency, but also of the academic disciplinary expectations and standards.

9.1 Principles of Data Management Framework

Key principles which underpin Data Management Frameworks are listed below:

i. Data management is necessary to support the ever evolving global research environment which is data-intensive;

ii. Data management is a necessary part of carrying out good research as well as supporting the wider global research community;

iii. Data management enables each researcher to use his/her data efficiently;

iv. Usually the data management framework of an individual institution falls within the existing external legal as well as regulatory frameworks such as the Australian Code for the Responsible Conduct among others;

v. Institutions involved in research usually support all aspects of the lifecycle of data such as collecting, storing, manipulating, sharing, collaboration, publishing, archiving and access for re-use;

vi. Collaboration and teamwork between researchers, information specialists, research administrators and technical support staff lead to achieving effective data management.

The Data Management Framework, as shown below, outlines some of the basic elements needed within an institutional context in order to support effective data management. The basic elements are set out in the following four separate categories:

- Institutional policy and procedures;
- IT infrastructure: the software, hardware and other facilities which support data-related activities;
- Support services: technical staff and other ways of providing training and advising as well as other forms of support, e.g. web-pages;
- Metadata management: for the data records to be used for internal as well as external purposes.

9.2 Variables and Indicators:

9.2.1 Variables

i. Definition:

A variable is a characteristic that changes from one member of a given (sample) population to another and it may assume a set of values to which a numerical measure can be assigned, which is more than one. Examples of variables include, for example, the height of class four students, the weight of patients who visit clinics/hospitals, the marital status of respondents, the sex of the patients and clients, who are the respondents.

ii. Description and Types:

The first two of these variables given in section (i) above (height and weight) are examples of quantitative (nominal) variables because they provide numerical information. The last two (marital status and sex) are qualitative (categorical) variables because they provide non-numerical measurement (information). This implies that variables can be classified into two categories: quantitative or qualitative. .

Quantitative variables:

These are variables which lend themselves to numerical measurement or treatment and can be further classified into discrete or continuous variables.

Discrete variables:

A variable is said to be discrete if it's possible values contain only numeric values such as 0, 1, 2, 3, 4… Examples of such variables might be the numbers of vehicles involved in road accidents on a certain road on different days, or the number of children in different families. These are both the results of enumeration and are, therefore, discrete variables. In other words, a variable is considered to be discrete, if it assumes values which are integers or a finite number of values.

Continuous variables:

Unlike a discrete variable, a continuous variable may reflect any value that lies between its maximum and minimum delimiting values, that is, it can assume an infinite number of values that belong to a set of real numbers. Quantities such as weight, length, or temperature, in principle, can be measured arbitrarily and accurately. The measurement may be restricted by the method employed, as well as by the accuracy of the measuring instrument. For example, the height of a child is a continuous variable because a child may be 1.23432985678902… meters tall. However, it is always the case that whenever a height of person is measured, it is usually done to the nearest centimeter as it is rounded off. In this illustration, the height of the child will be expressed as 1.23 meters or 1 meter and 23 centimeters.

9.3 Basic Statistics

i. Frequency Tables (Proportions)

Frequency tables are part of descriptive statistics and are used to describe sets of numbers, which are usually organized by researchers into tables and graphs which are referred to as frequency distributions. Consider example 1 below:

EXAMPLE 1

Students' scores on a math test were represented by the following numbers:

20,25,17,24,21,19,17,22,19,17,23,21,29,27

The frequency table below displays the distribution of a number of students, who achieved a specific score on the math test. In Example 1, a score of 17 was achieved by three students.

Math Score	Frequency	Percent	Summary Percentile
17	3	21.4	21.4
19	2	14.3	35.7
20	1	7.1	42.9
21	2	14.3	57.1
22	1	7.1	64.3
23	1	7.1	71.4

24	1	7.1	78.6
25	1	7.1	85.7
27	1	7.1	92.9
29	1	7.1	100.0
TOTALS	**14**	**100.0**	

Table 5: Frequency Table

The same information can be displayed in a frequency graph that shows the distribution or the number of students and the corresponding scores they achieved, as shown in the following bar graph:

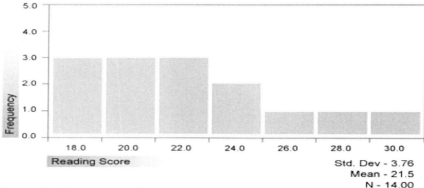

Figure 9: Frequency Graph

ii. Mean:

The arithmetic *'mean'* of a given variable is the sum of the values observed in the data divided by the total number of observations. It is a measure of the mid-point or centre of a number of quantitative variables and is an indicator, which is most commonly used to express a to have the variables, as well as the observed values of the variables represented by symbols so as to avoid assigning a fixed variable" to a specific set of numbers. Suppose we denote the given variable as x, then the ith observation in the given data set is represented by the symbol x_i. Suppose, the sample size is denoted by n, the mean of variable x:

$$\frac{x_1 + x_2 + x_3 + \cdots + x_n}{n}.$$

In order to simplify the sum in the numerator, the Greek symbol Σ, which is called sigma, is used. That is, the sum $x_1 + x_2 + x_3 \ldots + x_n$ in the numerator is denoted as

$$\sum_{i=1}^{n} x_i,$$

and read as "the sum of all x_i, with i ranging from 1 to n".
The sample mean $\bar{x}$ (bar x) can, therefore, be defined in terms of the symbols as the sum of the observed values $x_1 + x_2 + x_3 \ldots + x_n$ in a given data divided by the number of observations n, and is expressed as:

$$\bar{x} = \frac{\sum_{i=1}^{n} x_i}{n} \quad \text{or} \quad \frac{\sum x_i}{n}.$$

EXAMPLE 3:

Using the values in Example 1 above, the mean score is given by:

$$\bar{x} = \frac{20 + 25 + 17 + 24 + 21 + 19 + 17 + 22 + 19 + 17 + 23 + 21 + 29 + 27}{14}$$

$$= \frac{301}{14}$$

= 2.15 which is the mean value.

iii. Median

The value of the variable in a given dataset that divides a given set of observed values in half is the sample median of a variable, i.e. the median is the score found in the middle of a given frequency distribution (the 50th percentile score). In order to obtain the median value of a given variable, it is necessary to arrange the observed values in a given data set in ascending (increasing) order, after which the middle value can then be determined in the ordered list. In case there are an odd (as opposed to an even) number of observations, the sample median is taken to be the observed value, which is found exactly in the middle of the arranged (ordered) list. On the other hand, should there be an even number of observations, then the sample median is taken to be the number that is halfway between the two observed values in the arranged (ordered) list that are found in the middle. In other words, the values observed in one half of the data set are less than or greater than the value of the median and the observed values in the other half are greater than or equal to the value of the median. Thus, for both the odd and the even number of observations; if the number of the data set is denoted by n, then the position of the sample median is at *n*+1/2 in the arranged (ordered) list. In example 1, discussed above, the median score is 21.

iv. Percentile:

A percentile rank is the proportion of scores in a given distribution that is equal to or below a particular value. Percentiles divide a given set of data into one hundred (100) equal parts, unlike quartiles, which divide a given set of data into four equal parts. For instance, in example 1 discussed above, a score of 24 and below was achieved by 79% of the students. This implies that a score of 24 is found at the 79th percentile.

As discussed previously, the median of the sample variable usually divides the observed values of the data into two equal parts -- 50% at the bottom and 50% at the top. The percentile, on the other hand, divides the observed values into one hundred equal parts, that is, into hundredths. Thus, the first percentile, represented by P_1, is the number dividing the bottom 1% of all the observed values from the remaining top 99%. Similarly, the second percentile, represented by P_2, is the number dividing the bottom 2% of all the observed values from the remaining top 98%. Thus, the 50th percentile represents the median.

v. Confidence Level:

The confidence level is represented by the probability value $(1 - \alpha)$, which is associated with the confidence interval. It is usually expressed as a percentage. For instance, if $\alpha = 0.05 = 5\%$, then we will have a confidence interval of 95% as a result, i.e. $(1 - \alpha)$, = (1-0.05) = 0.95 = 95%.

For example, consider an opinion poll, which predicted that Democrat candidates would win 60% of all the votes cast, if the election were held today. Suppose the pollster attached a 95% confidence level, then the resulting confidence interval would be 60% plus or minus 3% (60% ± 3%). That is, the researcher believes that it is very likely that Democrat candidates would get between 57% and 63% of all votes cast

vi. Confidence Interval:

A confidence interval provides an estimated range of values, which are calculated from sample data, which is likely to include the population parameter which is unknown.

The main aim of a confidence interval is to portray the accuracy of what the sample's mean is believed to be. As more samples are taken, a decline occurs in the variance of the mean. It is expected that the measured mean is relatively more accurate as compared to the actual (population) mean. The assumption to be made is that the sample mean is normally distributed; although no assumption is made regarding the real distribution.

Suppose independent samples were repeatedly taken from the same population, and then from each sample a confidence interval was calculated, then a specific percentage (the confidence level) of those intervals would include the population parameter which is unknown. Confidence intervals are always calculated to give the percentage as 95%, although confidence intervals of 90%, 99%, or 99.9% can also be produced for the unknown parameters.

The uncertainty of the researcher about the unknown parameter is reflected by the confidence interval's width. In case of a very wide confidence interval, the indication is that there is the need for more data to be collected, before any definite conclusion can be drawn about the parameter.

Confidence intervals always provide more information than the much simpler results of a hypothesis test, in which a researcher has to make a decision on whether to reject null hypothesis (H_0) or not (null hypothesis is explained below). This is because confidence intervals provide a wider range of reasonable values for the unknown parameters.

vii. Significance Test

The term "significance", in everyday language, is defined as something that is important. The significance test is a statistical process employed by researchers to make a decision on whether to reject null hypothesis in favor of an alternate hypothesis. The significance test involves a comparison between the observed values and the hypothetical values. It establishes if a relationship exists between the variables of interest, or if the observed result is as a result of pure chance. The following steps should be followed when testing for statistical significance:

a) State the research hypothesis
b) State the null hypothesis
c) Select a probability of error level (α-level)
d) Select and compute the test for statistical significance
e) Interpret the results

For example, a researcher may find that a statistically significant relationship exists between the age of a citizen and their satisfaction level with a town's recreational services. It may be realized that citizens who are older, are 5% less satisfied with town recreational services than citizens who are younger. The 5% may seem to be a small difference, but that does not mean that it is not statistically significant.

viii. The Research Hypothesis:

A research hypothesis states the relationship that is expected between the two variables under consideration. The hypothesis may sometimes be stated in general terms, or may include dimensions of direction as well as magnitude.

ix. Null Hypothesis:

A null hypothesis often states that no relationship exists between the two variables of interest. For example, the strength of an individual is not influenced by age. A null hypothesis may also sometimes state that the relationship that is proposed in the research hypothesis is not true.

Null hypothesis is often used in research since it is much easier to reject a null hypothesis than to prove a research hypothesis. In other words, it is not very easy to show that something is always true, but it is much easier to show once that something is wrong. An alternative hypothesis is usually favored whenever a null hypothesis is rejected.

9.4 Sampling

Sampling is the act, technique or process of selecting a part of a statistical population to represent the entire statistical population for the purpose of determining characteristics or parameters of the entire statistical population (Johnson, 1992).

A statistical population is the record of qualitative traits or a set of measurements which correspond to the entire collection of particular units for which intelligent inferences are to be made. The statistical population usually represents the specific target of investigation.

A sample data set is the set of measurements that are drawn from a statistical population and which are actually collected during an investigation.

i. Sampling Frame

Sample frame is the set of units or individuals whose probability of being selected is greater than zero.

ii. Sampling Unit

A sampling unit is the element or a set of elements on which the actual measurement of a character is made, that is, it is the basic object upon which the experiment or the study is carried out.

In some cases, a study can involve units of a population at different levels simultaneously. This is referred to as multi-stage sampling. For instance, these stages may involve schools, classes and students, and the study may involve one or more of the levels which may be of scientific interest. For example the first possibility might be interest only in the students and the second possibility might be interest in schools and students simultaneously.

iii. Sample Size

Sample size is the number of units among the entire population selected for study.

iv. Sampling Methods

It is normally not practical to carry out a research study on a whole population of interest, particularly, if that population is too large to be reasonably interviewed in its entirety. This has made researchers consider sampling, whereby a portion of the population is selected in order to make a general inference about the entire population of interest. Thus, in sampling not all elements/members in the population of interest are investigated. So it reduces the number of elements/members to be studied, which in turn, leads to a reduction of the workload and the overall time and cost of executing the survey/study.

Various methods can used be in order to select a sample from a population as discussed below:

a. **Purposive:**

A purposive sample, commonly referred to as judgmental sample, is for a specific purpose or need. The selection of the sample is based on the purpose of the study and knowledge of the population. The researcher may have a particular target group in mind, for instance, the student leaders in a university, drug addicts who have successfully been rehabilitated, or legislators supporting a certain government bill. It may not be possible for the researcher to specify the population because all of its membership may not be known so accessing all of them may prove difficult. This may force the researcher to interview whoever is available by focusing on a specific target group.

Snowball sampling, which is a subset of purposive sampling, got its name from the manner in which a snowball accumulates snow. A snowball sample is obtained by requesting a participant to suggest another person, who might be appropriate or willing to participate in the study. This method of sampling is highly applicable in populations which are hard to track, such as drug users, truants, etc.

b. Simple random sampling

In simple random sampling, each subject is chosen wholly by chance, so, consequently, each subject in the population has equal probability of being selected. One of the most common criteria of obtaining a random sample from a population is to give each element/ member in the population a number, after which a table of random numbers is used to help in deciding those to be included.

c. Systematic sampling

In systematic sampling, elements/members are selected from a list of the entire population at regular intervals, which are selected in such a way that the sample size is adequate. For example, every 12th element in the population is selected to be part of the sample. Although this method may lead to bias, it is easy and convenient to use.

d. Probability sampling;

Probability sampling (representative samples) involves selecting samples in such a way that they represent the entire target population. The results they provide are more credible or valid because they reflect the characteristics that the target population from which they are selected bear, such as residents of a specific town, students at a college, etc. There are two types of probability samples: (1) random (2) stratified.

e. Clustered sampling

Subgroups of the target population in a clustered sample are used as sampling units, instead of the individuals. The population is first divided into subgroups, which are referred to as clusters. The clusters are then selected randomly so as to be included in the study. All elements/members in these selected clusters are included in the study. The fact that clustering was used should be taken into consideration when analysis of the data is being carried out. An example of cluster sampling is the General Household survey such as the one that is always undertaken in England annually. Households form the clusters and all the members of those selected households (clusters) are included in the survey.

Bias in Sampling

Whenever a researcher is selecting a suitable sampling method to use in his/her research study, the following potential sources of sampling bias should be taken into consideration, regardless of the sampling method used:

i. A bias can be introduced if there have been any changes from the pre-arranged sampling rules.

ii. A bias can be introduced which favors subject elements/members who live in hard to reach areas who might otherwise not be included in a purposive or random sample.

iii. A bias can be introduced if individuals who are initially selected are replaced with others, possibly because there is some physical or geographical difficulty in reaching those initially selected.

iv. A bias can be introduced if there is a low response rate. It is usually very important to maximize the likely rate of responses in a survey by over sampling.

v. A bias can be introduced if the list employed as a sample frame is out of date, which may result in the exclusion of individuals who have moved recently out of the study area.

9.5 Measurement Scales:

There are four main scales which are related to and build on each other, which include:

- Nominal (Categorical)
- Ordinal
- Interval
- Ratio

i. Nominal variable:

A nominal variable, which is sometimes referred to as a categorical variable, is a variable which has more than one category, although the categories may lack intrinsic ordering. Examples of categorical variables include gender, which is categorized as male and female, but has no intrinsic ordering. Another example worth mentioning is hair color which has a number of categories such as: blonde, brunette, brown, black, and red among others. There is no way to give an order to these categories except arbitrarily. A purely categorical variable allows for the assigning of the categories, but cannot in any clear way order the variables. Should the variable bear a clear ordering, it will fall under ordinal variable, which is described below.

ii. Ordinal variable:

An ordinal variable is, in many aspects, similar to a nominal or categorical variable. The main difference between them is that an ordinal variable has a clear order. For instance, economic status can be considered as a variable with three categories that is low, medium and high. If we take a variable such as educational experience with categories such as elementary school graduates, high school graduates and college graduates; these can further be ordered from lowest to highest as elementary school is followed by high school and then by college; although, the spacing between the values associated with them may not be similar across different levels of the variables. Suppose we assign the three levels of educational experience with scores 1, 2 and 3, and then the difference between categories one (elementary school) and two (high school) is compared to the difference between categories two (high school) and three (college). The difference existing between categories one and two is probably much bigger than the one existing between categories two and three. Thus, this example has shown mathematically that in an ordinal variable, the difference between two different categories varies that is, 2-1 ≠ 3-2. That is to say that people can be put in some order with respect to their level of educational experience, but there will be inconsistencies regarding the difference between the categories. If these categories have the same space between them; the variable would be considered as an interval variable as described below.

iii. Interval Variable:

Equal spaces exist between the values of interval variables. For example, suppose the annual income (measured in dollars) of some employees is considered as a variable, and four employees earn $20,000, $30,000, $40,000 and $50,000 respectively. The size of interval between the earnings of the first and the second employees is $10,000. The same interval exists between the difference in earnings between the second and the third employees as well as between the third and fourth employees. Thus, in each of those cases, the interval is the same ($10,000). Suppose two other high ranking employees earn $210,000 and $220,000 respectively, then the size of the interval between their earnings is also $10,000.

iv. Ratio variable:

Numbers can be meaningfully compared as numerical multiples of one another in ratio variable. It is important to note that zero as a number has a meaning in a ratio variable. For example, we can compare the weight of a father and a son, and conclude that the father is twice as heavy as the son. That is, the father may weigh 86 Kg, whereas the son weighs 43 Kg. -- the difference between the weight of the father and the son is 43 Kg, which is the same as the difference between the weight of two individuals who weigh 90 Kg and 47 Kg. Multiplication as well as division are two important operations that can be used in ratio data, not only because the difference between 1 and 2 is equivalent to the difference between 3 and 4, but, more importantly, if one doubles 2 one gets 4. The similarity between ratio data and interval data is that both are quantitative, that is, they both measure quantities.

9.5.1 *The Importance of Choosing an Appropriate Scale*

Statistical analyses and computations assume that there are specific levels of measurement associated with the variable. For instance, computing the average hair color does not make sense. So it is clear that it does not make sense to compute an average of a categorical variable because categorical variable has no intrinsic ordering of levels in the categories. A result which does not make sense would also be obtained if the average of an ordinal variable such as educational experience computed. The meaning of such an average would be questionable because of the uneven spacing that exists between the three levels of education as discussed above in the ordinal section. It is necessary for a variable to have interval for its average to be calculated.

Some variables exist which are considered to be falling in between interval and ordinal. For example, if we consider a five-point Likert scale, where values 1, 2, 3, 4 and 5 are assigned to "strongly disagree", "disagree", "neutral", "agree" and "strongly agree" respectively; there is no certainty that the intervals between each of the five values above are the same (or equally spaced). So we must eliminate the possibility of the variable being an interval variable, but instead regard it as an ordinal interval. It is, therefore very important to assume that there are equal spaces between the intervals in order for one to use statistics which assume that the variable is interval.

9.6 Data and Information

i. Primary data

Primary data refers to data that is obtained or collected from first-hand experience such as data collected as the result of a face-to-face interview with a respondent as well as the data obtained by a researcher through experiment, focus group deliberations, questionnaires, surveys and by taking measurements.

ii. Secondary data

Secondary data refers to data that was obtained from another party or was gathered in the past. Secondary data is readily available and can often be obtained publicly from journals, publications and newspapers.

iii. Qualitative data

Quantitative data includes anything that can be expressed numerically (such as a number) or can be quantified. Quantitative data may be represented by interval, ratio or ordinal scales and is often used in most statistical manipulations. Examples of qualitative data include height of a child, score of students in a test, and number of students in a class.

iv. Quantitative data

Qualitative data can be observed, but cannot be measured and deals in large part with descriptions. Thus, it is not possible to express qualitative data as a number. Qualitative data include the data which represents nominal scales, e.g., gender, religious preference, and socio-economic status, among others.

9.6.1 *Organization of the Data*

Data can be obtained by making observations on the values of the variables for one or more objects. Each of the pieces of data is referred to as an observation and, in extension; the collection of all possible observations for a specific variable is referred to as a data matrix or data set. A data set (data matrix) consists of the total value of all variables, which are obtained and recorded for a particular set of sampling units.

In most cases, the values involving qualitative variables are usually coded by having numbers assigned to the different categories for easier manipulation (involving recording and sorting) thus by having the categorical data converted to numerical data, albeit in a trivial sense. For instance, the marital status of a person may be categorized as married, single, divorced, or widowed and the categories may be coded as 1, 2, 3, and 4 respectively. The coded data will still be regarded as nominal data. It should be well noted that the coded numerical data does not, in any way, share any of the properties associated with the numbers in the ordinary arithmetic sense. Therefore, it is not advisable to use the codes given above to write $4 > 2$ or $1 < 3$, just as we can also not write $4 - 2 = 3 - 1$ or $1 + 2 = 3$. This shows the importance of checking the legitimacy of the mathematical treatment of specific statistical data.

Data is usually presented in the form of a matrix (data matrix). Every value of a specific variable is organized in the same column and forms a column in the matrix; whereas, the observation (measurement which is collected in the sampling unit) forms a row in the matrix. Suppose the number of variables is k, and the number of observations (sample size) is n, then the resulting data set will look as follows:

$$\text{Sampling units} \begin{pmatrix} x_{11} & x_{12} & x_{13} & \cdots & x_{1k} \\ x_{21} & x_{22} & x_{23} & \cdots & x_{2k} \\ x_{31} & x_{32} & x_{33} & \cdots & x_{3k} \\ \vdots & & & \ddots & \\ x_{n1} & x_{n2} & x_{n3} & \cdots & x_{nk} \end{pmatrix}$$

where x_{ij} is a value of the jth variable collected from i^{th} observation, $i = 1, 2, \ldots, n$ and $j = 1, 2, \ldots, k$.

9.7 Statistical Software Packages

In most social as well as market research, statistical software is expected to do the following:

- Data: getting (typing/reading) data into a computer
- Data analysis: processing the data
- Data output: articulating and interpreting data input into the computer in a form which can be understood by the intended audiences.

The three steps may be expanded to include the following:

- Organize data
- Compare data
- Manage data
- Create new variables
- Summarize data (transform raw data)
- Generate tables and graphs

Some of the common statistical software packages include:

- Epi Info
- SPSS
- SAS
- STATA
- Systat
- NVivo

9.7.1 Epi Info

Epi Info is statistical software that is oriented towards applications associated with public health. Its analytical capabilities are limited to proportions, means, and risk ratios as well as risk differences, odd ratios, and 2-way cross tabulations. It can also be run using syntax just like STATA and SPSS, although it also includes GUI (graphical user interface), which enables the analysis to be menu driven.

9.7.2 SPSS

SPSS is one of the most widely used software packages/programs for statistical analysis in social science. It is used by health researchers, market researchers, survey companies, education researchers, government experts, and marketing organizations, among others. Other than statistical analysis, SPSS can also be used in data management (including file reshaping, case selection, and creating derived data), as well as in data documentation (such as a metadata dictionary which is stored with the data).

9.7.3 SAS

Statistical Analysis System (SAS) is software based on SAS programs which define a sequence of operations, which are to be performed on data stored in the form of tables. Although SAS contain graphical user interfaces (GUI), such as the SAS Enterprise Guide, most of the time, these graphical user interfaces (GUI) are just a front-end to facilitate or automate generation of SAS programs. SAS components usually expose their functionalities through application programming interfaces, which exist as statements and procedures.

9.7.4 STATA

STATA, versions 7.0 and higher, are capable of performing various statistical procedures on sample survey data which is complex. STATA also has graphics capabilities and be run using syntax just like SAS. STATA also has graphical user interface (GUI), which enables the analysis to be menu driven.

The table below compares the analytic capabilities of the four statistical software packages for analysis of complex survey data.

Estimate or Analysis	SAS 9.2 and higher	STATA 12 and Higher	SPSS 19 and higher	Epi Info 3.5.3
Means, proportions, Cross tabulations	X	X	X	X
Totals	X	X	X	
Ratios	X	X	X	
Median and other	X			

percentiles				
Odds ratios, risk ratios	*	X	X	X
Odds differences				X
Test for Difference in domain Means	**	X	**	**
Chi-square tests	X	X	X	
Linear regression	X	X	X	
Logistic regression	X	X	X	
Polychotomous logistic regression	X	X	X	
Survival analysis	X	X	X	
Poisson regression		X		
Additional regression models		X		
Design effect	X	X	X	X

Table 6: Analytic capabilities of four statistical software packages for analysis of complex survey data (YRBSS, 2014)

* In SAS, odds ratios can be obtained using logistic regression.

** In SAS and SPSS, a two-sample test for comparing means can be obtained using linear regression. Epi Info provides an estimated difference in domain means with standard error and confidence intervals on the differences provided.

10. Student Test (t-test)

The **t distribution** (also known as **Student's t-distribution**) is a common probability distribution, which is often used to estimate population parameters when dealing with a small sample size and/or when the population variance is unknown. When dealing with a sample size which is significantly large, the sampling distribution of a statistic (e.g. sample mean) will follow a normal distribution. Thus, when the standard deviation of a population is known (for a large sample), a z-score can be computed and then the normal distribution can be used to work out the probabilities with the sample mean.

$$Z = \frac{\overline{x} - \mu_{\overline{x}}}{\sigma_{\overline{x}}} = \frac{\overline{x} - \mu_0}{\sigma/\sqrt{n}}$$

In the case of t distribution, its values are given by:

$$T = \frac{\overline{X} - \mu}{S/\sqrt{N}}$$

Where μ is the population mean, $\overline{x}$ is the sample mean, S is the estimator for population standard deviation and N is the sample size.

It is important to note that the distribution of the t statistic is referred to as t distribution. The importance of t distribution is that it makes it possible to conduct statistical analyses on data sets that are usually not appropriate for analysis using normal distribution.

10.1 Survey Data Analysis in Stata

Example:

Real-world, publicly available survey data is often very complex (see the DHS example).

Consequently, we will contrive an example for this tutorial, estimating p, the prevalence of a disease, say malaria, in a hypothetical country, called "Inventia".

Country profile:

Province	Population size	Number of districts
1	225,000	50
2	150,000	42
3	100,000	32
4	25,000	23
Total	500,000	146

In Inventia, the climate differs between provinces. For instance, Province 4 is more arid and at a higher altitude than the rest of the country. Consequently, the prevalence of malaria p varies between provinces. Also, access to malaria prevention is not consistent across the country, subsequently p may also vary somewhat between districts. (For instance, urban populations may have more resources to prevent malaria, so thus a lower prevalence.)

The true prevalence of malaria in Inventia is 13.1%.

Here, we review how to analyze data from several different survey designs:

_ **Simple Random Sampling** - We randomly sample 1,000 people from Inventia.

_ **Stratified Sampling** - We randomly sample 250 people from each of the 4 provinces of Inventia.

_ **Cluster Sampling** - We randomly sample 25 districts from Inventia and randomly sample 40 people within each district.

_ **Stratified Cluster Sampling** - For each of the 4 provinces, 5 districts would be randomly sampled and 50 people sampled within these 20 districts.

10.2 Analyzing Survey Data in Stata

In order to analyze survey data in Stata, one must first "svyset" your data. This command tells Stata what survey design was used to obtain the data. This includes a specification of survey weights, the finite population correction(s), and levels of clustering and stratification. Once Stata has this information, it incorporates the specified design elements into its calculations. One can then use the survey estimation procedures in Stata. For example, svy: meanvar name, svy: proportion var name, svy: regress ....

Before analyzing the survey data, one must be able to answer the following questions:

1. What is the design of my survey?
2. Am I using a finite population correction? At which stage of the design?
3. What are the survey weights used in the design?

Once you know these things, you can start analyzing your data in Stata.

10.2.1 Simple Random Sampling

Design: We randomly sample 1,000 people from the entire country of Inventia.

Notation - STATA:

_ N is the total population size

_ n is the number of individuals sampled from the population without replacement

In our case, n = 1; 000 and N = 500; 000.

Finite Population Correction: $1 - f = (1 - \frac{n}{N})$

Survey Weights $w_i = P(\text{individual } i \text{ is included in the survey})^{-1} = \frac{N}{n}$

Exercise: Estimate the prevalence of malaria in Inventia.

```
use "srs.dta", clear

generate weight_srs = pop_size/1000
generate fpc = 1000/pop_size * note that this does not match the definition above
svyset id [pweight=weight_srs], fpc(fpc)
svy: proportion malaria

svyset id [pweight=weight_srs]
svy: proportion malaria
estat effects, deff
```
a

proportion malaria

Why does it not matter much if the finite population correction is used in this example?

Exercise: Estimate the prevalence of malaria in each of the four provinces.

svy, sub(if province==1): proportion malaria

svy, sub(if province==2): proportion malaria

svy, sub(if province==3): proportion malaria

svy, sub(if province==4): proportion malaria

Is there evidence of province-level variation in malaria prevalence?

10.2.2 Stratified Sampling

Design: We randomly sample 250 people from each of the 4 provinces of Inventia.

Notation:

_ N is the total population size

_ Nj is the population in province j, j = f1; 2; 3; 4g.

_ nj individuals are sampled from province j.

The important design question in stratified sampling is how to choose the sample size within each stratum. In our case, N1 = 225; 000;N2 = 150; 000;N3 = 100; 000 and N4 = 25; 000.

nj = 250 for each j.

Finite Population Correction: $1 - f_j = \left(1 - \frac{n_j}{N_j}\right)$

Survey Weights: $w_{ij} = P(\text{individual } i \text{ in strata } j \text{ is in the survey})^{-1} = \frac{N_j}{n_j}$

Exercise: Estimate the prevalence of malaria in Inventia.

```
use "stratified.dta", clear

proportion malaria
proportion malaria, over(province)
generate weight_stratified = prov_size/250
generate fpc_stratified = 1/weight_stratified
svyset id [pweight=weight_stratified], strata(province) fpc(fpc_stratified)
svydescribe weight
svy: proportion malaria
estat effects, deff
```

10.2.3 Cluster Sampling

Design: We randomly sample 25 districts (clusters) from Inventia; within each district, we randomly sample 40 people.

Notation:

_ N is the total population size

_ Nk is the population size in district k, k = f1; ::::; 146g

_ nI out of NI total districts are sampled for inclusion in the survey (primary sampling unit)

_ We sample nk individuals in district k and select them for inclusion in the survey (secondary sampling unit)

In our survey, nI = 25, NI = 146, nk = 40, and Nk is the population size in district k.

10.2.4 Stratified Cluster Sampling

Finite Population Correction:
Stage I: $1 - f_I = \left(1 - \frac{n_I}{N_I}\right)$
Stage II: $1 - f_k = \left(1 - \frac{n_k}{N_k}\right)$

Survey Weights':

$$w_{ik} = P(\text{individual } i \text{ in cluster } k \text{ is in the survey})^{-1}$$
$$= [P(\text{cluster } k \text{ selected}) * P(\text{individual } i \text{ in cluster } k \text{ selected } | \text{ cluster } k \text{ selected})]^{-1}$$
$$= \frac{N_I}{n_I} * \frac{N_k}{n_k}$$

Exercise: Estimate the prevalence of malaria in Inventia, using only the first stage finite population correction.

```
use "cluster.dta", clear

generate fpc1 = 25/146
generate fpc2 = 40/districtsize
generate weight_cluster = (fpc1*fpc2)^-1
svyset district [pweight=weight_cluster], fpc(fpc1) || id, fpc(fpc2)
svy: proportion malaria
estat effects, deff
```

We could combine stratified, cluster and simple random sampling all into one design!

Design: For each of the 4 provinces, we randomly sample 5 districts. Within each of the 20 districts, we randomly sample 50 people.

Survey weights: As an example, for province 2:

$P(\text{person } i \text{ in district } j \text{ in province 2 in survey})$
$= P(\text{district } j \text{ in survey} | \text{province 2}) P(\text{person } i \text{ in survey} | \text{district } j)$
$= \frac{5}{42} * \frac{50}{districtsize_j}$

Finite population correction:

Stage I: $\frac{\#\text{sampled districts}}{\text{total }\#\text{ districts in the province}}$

Stage II: $\frac{\#\text{sampled per district}}{\text{district population}} = \frac{50}{districtsize_j}$ for district j.

Exercise: Estimate the prevalence of malaria in Inventia.

```
use "stratifiedcluster.dta", clear

generate fpc1 = 5/ndistrict
generate fpc2 = 50/districtsize
generate weight_stratcluster = (fpc1*fpc2)^-1
svyset district [pweight=weight_stratcluster], fpc(fpc1) strata(province) || id, fpc(fpc2)
svy: proportion malaria
estat effects, deff
```

Confidence intervals with the t-distribution in Stata

Suppose t is a random variable that follows a t-distribution with n degrees of freedom.

`tden(n,t)`	returns the probability density function of Students t distribution
`ttail(n,t)`	returns the reverse cumulative (upper tail or survivor) Students t distribution
`invttail(n,p)`	returns the inverse reverse cumulative (upper tail or survivor) Students t distribution

Note that if `ttail(n,t)`= p, then `invttail(n,p)` = t.

Stata will calculate confidence intervals for you:

Calculator: cii n mean sd, level(95)

Function: ci varlist, level(95)

There is no Stata function for calculating confidence intervals for normally distributed data when the standard deviation is known, since this scenario doesn't really happen in practice.

1. Calculate the mean and standard deviation of BMI at baseline.

   . summarize bmi1

2. Take a sample of size 20 from the Framingham cohort. Calculate the mean and

   standard deviation of BMI at baseline in the subsample (I use set seed 2, if you want

   to get the same sample as me). We are interested in making inferences about BMI at

   baseline in the total Framingham cohort using only the sample of size 20.

   . set seed 2

   . drop if bmi1 == .

   . sample 20, count

   . sum bmi1

3. Assume that the sample standard deviation is known (and equal to the standard deviation in the Framingham cohort). Construct a 95% confidence interval for the mean BMI in your subsample. Note that if normal (z)= p, then invnormal(p) = z:

   95% CI: _x _ Z0:975 σ/√n

   . di 25.0 - invnormal(0.975)*4.1/sqrt(20)

   . di 25.0 + invnormal(0.975)*4.1/sqrt(20)

4. Use invttail to construct a 95% confidence interval for the mean BMI in your sub sample σσσ by hand, now assuming that the sample standard deviation is unknown.

    . di 25.0 - invttail(19, 0.025)*3.2/sqrt(20)
    . di 25.0 + invttail(19, 0.025)*3.2/sqrt(20)

5. 5. Use cii to construct a 95% confidence interval for the mean BMI in your subsample.

    . cii 20 25.0 3.2

6. Use ci to construct a 95% confidence interval for the mean BMI in your subsample.

    . ci bmi1

Two Sample t-tests in Stata

Example: In the Framingham cohort, we want to examine the distribution of heart rate at exams 1 and 2. Specifically, we wish to test whether there is a difference in mean heart rate between exam 1 and exam 2. Additionally, we are interested in whether the mean heart rate differs between men and women at exam 2. We sample 100 people from the Framingham cohort.

For this example, use the dataset heartrate.dta on this webpage, which contains the random sample of 100 participants.

Hypothesis testing with paired data in Stata:

```
. ttest heartrte1 == heartrte2

Paired t test
------------------------------------------------------------------------------
Variable |     Obs        Mean    Std. Err.   Std. Dev.   [95% Conf. Interval]
---------+--------------------------------------------------------------------
heartr~1 |     100       75.03    1.290247    12.90247    72.46987    77.59013
heartr~2 |     100       76.17    1.293031    12.93031    73.60435    78.73565
---------+--------------------------------------------------------------------
    diff |     100       -1.14    1.344125    13.44125    -3.807035    1.527035
------------------------------------------------------------------------------
     mean(diff) = mean(heartrte1 - heartrte2)                 t =  -0.8481
 Ho: mean(diff) = 0                           degrees of freedom =       99

 Ha: mean(diff) < 0          Ha: mean(diff) != 0         Ha: mean(diff) > 0
  Pr(T < t) = 0.1992       Pr(|T| > |t|) = 0.3984        Pr(T > t) = 0.8008
```

```
. gen hdiff = heartrte2 - heartrte1

. ttest hdiff== 0

One-sample t test
------------------------------------------------------------------------------
Variable |     Obs        Mean    Std. Err.   Std. Dev.   [95% Conf. Interval]
---------+--------------------------------------------------------------------
   hdiff |     100        1.14    1.344125    13.44125    -1.527035    3.807035
------------------------------------------------------------------------------
    mean = mean(hdiff)                                          t =   0.8481
Ho: mean = 0                                    degrees of freedom =       99

   Ha: mean < 0              Ha: mean != 0              Ha: mean > 0
 Pr(T < t) = 0.8008     Pr(|T| > |t|) = 0.3984      Pr(T > t) = 0.1992
```

The commands `ttest heartrte2 == heartrte1` and `ttest hdiff==0` lead to the same test.

This command can be found through the following drop-down menus: Statistics / Summaries, tables, and tests / Classical tests of hypotheses/ Mean-comparison test, paired data.

Hypothesis testing with unpaired data and equal variances in Stata:

```
. ttest heartrte2, by(sex1)

Two-sample t test with equal variances
------------------------------------------------------------------------------
   Group |    Obs        Mean     Std. Err.   Std. Dev.   [95% Conf. Interval]
---------+--------------------------------------------------------------------
    Male |     39     76.82051    2.042025    12.75244    72.68665    80.95438
  Female |     61     75.7541     1.681246    13.13095    72.39111    79.11709
---------+--------------------------------------------------------------------
combined |    100     76.17       1.293031    12.93031    73.60435    78.73565
---------+--------------------------------------------------------------------
    diff |            1.066414    2.662326               -4.216884    6.349713
------------------------------------------------------------------------------
    diff = mean(Male) - mean(Female)                              t =   0.4006
Ho: diff = 0                                     degrees of freedom =       98

    Ha: diff < 0                 Ha: diff != 0                 Ha: diff > 0
 Pr(T < t) = 0.6552         Pr(|T| > |t|) = 0.6896          Pr(T > t) = 0.3448
```

Hypothesis testing with unpaired data and unequal variances in Stata:

```
. ttest heartrte2, by(sex1) unequal

Two-sample t test with unequal variances
------------------------------------------------------------------------------
   Group |    Obs        Mean     Std. Err.   Std. Dev.   [95% Conf. Interval]
---------+--------------------------------------------------------------------
    Male |     39     76.82051    2.042025    12.75244    72.68665    80.95438
  Female |     61     75.7541     1.681246    13.13095    72.39111    79.11709
---------+--------------------------------------------------------------------
combined |    100     76.17       1.293031    12.93031    73.60435    78.73565
---------+--------------------------------------------------------------------
    diff |            1.066414    2.645081               -4.194674    6.327503
------------------------------------------------------------------------------
    diff = mean(Male) - mean(Female)                              t =   0.4032
Ho: diff = 0                     Satterthwaite's degrees of freedom =  82.8637

    Ha: diff < 0                 Ha: diff != 0                 Ha: diff > 0
 Pr(T < t) = 0.6561         Pr(|T| > |t|) = 0.6879          Pr(T > t) = 0.3439
```

This command can be found through the following drop-down menus: Statistics / Summaries, tables, and tests / Classical tests of hypotheses/ Two-group mean-comparison test.

Instead of the data structure above, suppose that, in your dataset, you have heart rate for men in one variable/column and heart rate for women in another variable/column (instead of our situation where we have heart rate in one variable and sex as another variable). How do you perform a t-test then? Use the command ttest heartratew == heartratem, unpaired unequal, where heartratew is the heart rate variable for women and heartratem is the heart rate for men. It is important to use the option unpaired.

If you do not use this option, Stata will perform a paired t-test. You may also choose leave out the unequal option if you wish to assume equal variances.

The following 4 lines of code transform the data to the situation where we have heart rate for men in one variable (heartrtem) and heart rate for women in another variable (heartrtew). It is not necessary to memorize or understand this portion of code. It is simply included for completeness. The fifth line of code runs the two sample t-test.

```
. gen id = _n
. reshape wide heartrte2, i(id) j(sex1)
. rename heartrte21 heartrtem
. rename heartrte22 heartrtew

. ttest heartrtew = heartrtem, unpaired unequal

Two-sample t test with unequal variances
------------------------------------------------------------------------------
Variable |    Obs        Mean     Std. Err.  Std. Dev.  [95% Conf. Interval]
---------+--------------------------------------------------------------------
heartr~w |     61     75.7541     1.681246   13.13095    72.39111    79.11709
heartr~m |     39    76.82051     2.042025   12.75244    72.68665    80.95438
---------+--------------------------------------------------------------------
combined |    100       76.17     1.293031   12.93031    73.60435    78.73565
---------+--------------------------------------------------------------------
    diff |             -1.066414  2.645081              -6.327503    4.194674
------------------------------------------------------------------------------
    diff = mean(heartrtew) - mean(heartrtem)                 t =  -0.4032
Ho: diff = 0                          Satterthwaite's degrees of freedom = 82.8637

    Ha: diff < 0                 Ha: diff != 0                 Ha: diff > 0
 Pr(T < t) = 0.3439         Pr(|T| > |t|) = 0.6879         Pr(T > t) = 0.6561
```

This command can be found through the following drop-down menus: Statistics / Summaries, tables, and tests / Classical tests of hypotheses/ Two-sample mean-comparison test.

10.2.5 ANOVA in Stata

In this example, we will use data from the California Health Interview Survey (CHIS). From their website (http://www.chis.ucla.edu): CHIS is the nation's largest state health survey. Conducted every two years on a wide range of health topics, CHIS data gives a detailed picture of the health and health care needs of California's large and diverse population. CHIS is conducted by the UCLA Center for Health Policy

Research in collaboration with many public agencies and private organizations.

In 2009, CHIS surveyed more than 47,000 adults, more than 12,000 teens and children and more than 49,000 households. We will use a sample of 500 adults for this lab (CHISANOVA.dta).

Suppose we are interested in the relationship between the number of hours worked (per week) and health, as measured by BMI. Would we expect those who worked longer hours to be healthier than those who worked shorter hours, or vice versa? Number of hours worked per week is divided into 5 categories: 0-10, 10-25, 25-35, 35-45, 45+.

1. How many people are in each category?
2. We now wish to run an ANOVA. Are the assumptions for ANOVA met?
3. What are the null and alternative hypotheses for this test?
4. Perform the hypothesis test at the _ = 0:05 level.

Conduct a oneway ANOVA in Stata using the oneway command:

Conduct a oneway ANOVA in Stata using the oneway command:

```
. oneway bmi work_cat, tabulate

             |         Summary of bmi
    work_cat |       Mean    Std. Dev.       Freq.
-------------+------------------------------------
        0-10 |   26.431579    5.9410147         38
       10-25 |   26.429189    5.7075504         74
       25-35 |     24.3495    4.1477871         60
       35-45 |   27.128351     5.647101        188
         45+ |   27.854928    6.1797228        140
-------------+------------------------------------
       Total |     26.8419    5.7540637        500

                        Analysis of Variance
    Source              SS         df      MS             F     Prob > F
------------------------------------------------------------------------
Between groups      550.823688      4   137.705922       4.27    0.0021
 Within groups      15970.6916    495   32.2640234
------------------------------------------------------------------------
    Total           16521.5153    499   33.1092491

Bartlett's test for equal variances:  chi2(4) =  11.7543  Prob>chi2 = 0.019
```

You may also use the following drop-down menus to access the oneway command: Statistics/ Linear models and related / ANOVA/

MANOVA / One-way ANOVA.

What are:

(a) your test statistic,
(b) the degrees of freedom,
(c) the p-value,
(d) your decision, and
(e) your interpretation

5. We have rejected the null hypothesis, thus we have evidence that at least one pair of means are not equal. Perform all possible pairwise comparisons using the Bonferroni correction.

6. Which pairs of means are significantly different?

7. A colleague of yours, who has the same dataset, calculates the means for each work category. After looking at these means, he takes the group with the largest mean (45+) and the group with the smallest mean (25-35) and performs a t-test (without a Bonferroni correction). He tells you that since he only did one test, he does not need to correct for multiple comparisons and that his method is valid. Do you agree? Why or why not?

10.2.6 Real Life Case Study (t-test)

Here is a real life example of an evaluation conducted by the author in one African country. The objective of the evaluation was to find out if the effectiveness of additional contraceptive logistics management training improved client load of facilities in the treatment arm. In other words, were facilities with additional training in commodity logistics management likely to perform better than those in other facilities where no additional training was provided? With regard to the strategic framework illustrated in the M and E section and with regard to the objective of the evaluation, one realizes right away that the emphasis is on a result at the higher -- in this case at the outcome level.

10.2.6.1 Study Design

The clinics included in this evaluation process were selected for their location (urban & rural), their provision of Family Planning services and the frequency of service availability, work and client load, and accessibility. An attempt was made to choose a balanced number of rural and urban clinics, when possible, making the selection process purposive.

The final study design product was an identification of two circuits or arms: one in which providers were trained (treatment clinics) and one where no formal training was conducted (control). 27 treatment clinics and 25 controls were selected for this study. The evaluation process was conducted in May 2001, roughly one year after the initial training.

10.2.6.2 Survey Methodology

In an ideal setting, a double-blind case/control study would have been more appropriate. For practical, logistical and economic reasons, it was not possible to implement these activities based on the traditional case/control study protocols. The experimental design has therefore been limited to a quasi-experimental design [Campbell, D. and Stanley, J.].

Analysis

The mean number of and median number of clients, as recorded in the service statistics (users of family planning (FP) methods) and the logistics data for family planning methods for the two circuits were then compared using t-tests and a corresponding confidence level of 0.05. In general, there were at least twice as many clients served every month in the treatment arm as compared to the control arm during February, March and April.

The average number of oral contraceptive clients, in Figure 10 illustrates a significant difference between the two arms over a three-month period. Health facilities in the treatment circuit served on the average 37, 52 and 44 clients respectively during the months of February, March and April with respective p-values of .005, .004, .004. Similar means for the control arms were 15, 13 and 17. While the control circuit averages remained generally stable, there was a range of between 37 and 52 among the treatment health facilities with a peak in March. Between February and March, the treatment centres served over twice the amount of people than the control arm health facilities did in February, and over three times in both March and April. The data suggests that FP facilities in treatment circuits served more clients on average than in the control arm, which indicates that the training was effective. In statistical terms, the hypothesis will be as follows: the two means in the two sets of facilities were different. The conclusion is that there is not sufficient evidence to reject the null hypothesis. Hence, if we were to perform this exercise in the communities, 19 out of 20 times, the treatment circuits would outperform the controls.

11. Data Management Model

Introduction

The data management model is a set of procedures through which information is processed. It is concerned with the collection of data, the manipulation of data, the storage and the retrieval of information. The main purpose of the data management model is to ensure that high quality data is obtained to make sure that the variabilities observed within the collected data originate from the phenomena being studied and not from the process of data collection. The other important purpose of the data management model is to ensure appropriate, accurate, and defensible analysis as well as a valid interpretation of the data.

Case Study: See the case study in Appendix 2b

11.1 Conceptual Framework

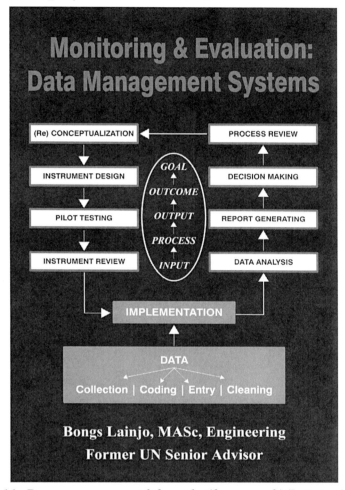

Figure 11: Data management life cycle (framework)(Lainjo, 2013)

The data management life cycle (framework) is made up of the following components:

- ➢ Conceptualization
- ➢ Instrument design
- ➢ Pilot testing
- ➢ Instrument review and finalization
- ➢ Implementation

- Data Analysis
- Report Generation
- Decision Making
- Process review

11.1.1 *Conceptualization*

Conceptualization is the process in which constructs are given theoretical meaning. It involves the definition of the constructs (or concepts) abstractly in theoretical terms. It is necessary to operationalize the concept(s) when testing hypotheses as well as when describing social phenomena.

Conceptualization includes:

i. Problem Identification

The problem to be solved by the research project must be well thought out as it helps identify the important concepts that represent the phenomenon being studied.

ii. Problem Definition

Research begins with a topic or a problem to be solved. It is necessary to have the problem well defined to make it easy to comprehend.

iii. Input Parameters

The parameters to be keyed in should be precise.

iv. Output Parameters

The processed parameters should be decoded in a language that can easily be understood by the target audience.

v. Expert Consultation

It is usually necessary to consult an expert to help provide the important information that may be pertinent in the research process.

vi. Constraints Identification

Factors that may restrict or limit financial, physical and (or) social aspects of the research project should be identified and the necessary steps taken to overcome them.

vii. Budget Preparation

A comprehensive budget should be prepared to include manpower, materials, transport, training, security and other logistical considerations.

viii. References

The materials such as books, journals, etc. used for consultation when writing the literature of the research project should be well referenced.

11.1.2 *Instrument Design*

i. **Instrument Type:**

- **Census**

    A census is the gathering of diverse information about a population or a simple count of the elements in a population.

- **Survey**

    Statistical surveys are basically the general methods which are used to gather quantitative information about a specific population. Population in this context is not limited to human beings/animals, but may be made up of other types of units such as households, firms, hospitals, herds, family units, etc. Examples of surveys include opinion polls, monitoring customer perceptions and public opinion, etc.

- **Cross sectional**

    Cross sectional studies, also referred to as transversal studies, are observational studies involving data collection from a population (or a sample) at one specific point in time.

- **Longitudinal**

    A longitudinal study is an observational study that involves data collection from the same population (or sample) repeatedly over a defined period of time (week, month, quarter, or year).

- **Routine Reporting (C/F prism)**

    A routine report of a research study can be done regularly usually over a specific duration such as daily, weekly, monthly or quarterly.

ii. **Identification of Major Components**

iii. **Identification of Minor Components**

iv. **Determination of Relationships**

v. **Determination of Format**

- ☐ Quantitative
- ☐ Qualitative
- ☐ Other

vi. **Identify layout structure**

vii. **Identify consistency checks**

viii. **Flow and skip patterns**

11.1.3 *Pilot Testing*

Pilot testing basically means a small-scale prior study or experiment where a pre-determined, but limited number of people (examinees) are identified, contacted and asked to complete the survey as well as to comment on various mechanics of the exercise. This enables the researcher to be aware if a key informant interview guide, observation form or survey is practical and can operate without contradiction in the real world. The most important aspects of pilot testing include:

i. Data recorder training
ii. Instrument screening
iii. Neutral site identification
iv. Community mobilization
v. Determination of pilot period
vi. Logistics
vii. Pilot implementation.

11.1.4 *Implementation*

i. Identify mapping features
ii. Produce guidelines on ethical issues
iii. Produce daily data flow
- Pre survey processes
- Survey processes
- Post survey processes
iv. Produce logistical guidelines
v. Produce a timeline
vi. Review input processes
vii. Review output processes

11.1.5 *Data Collection*

i. Teams -- Identify team
- Leaders
- Supervisors
- Editors
- Evaluators
- Interviewers/census takers
ii. Orient team
iii. Produce reporting guidelines
iv. Produce data collection guide
v. Produce data code book
vi. Utilization of electronic kits
vii. Data types
- Metric
- Ordinal
- Nominal
viii. Selection of clerks
ix. Task distribution allocation list
x. Collating

xi. Categories of entry
- ☐ Single
- ☐ Double

xii. Error rate

xiii. Back up frequency

xiv. Screen design

xv. Supervision

xvi. Break intervals

11.2 Teams

The definition of team in this context is comprised of a primary or core group and a number of subgroups. The latter are sub-sets of the former. These groups AS A RULE are made up of odd numbers. For example, if there is only one expert available, a team will not be possible. If there are two experts available, then one group can be established and where there is any disagreement whether to recommend an indicator or not, a coin is tossed (or flipped). Furthermore, if three experts are available, a subgroup will represent a group. That all the three members will work as a group and the recommendations will be considered a group decision based on degree of concordance. The process in this case is simple and straight forward. That is the majority decision (in this, two out of three) prevails. That is how the rule of odd numbers is used. (the preceding description addresses outlier scenarios).

As far as possible, this model works better if as many subgroups and groups as possible can be established without losing sight of the distribution (odd number of subgroup and group members), but usually, there are not more than five. For example, if there are ten experts, two subgroups of five members may be designated. In this case, the core group membership will be ten, while the subgroups will be two. In general, the total number of subgroup members should be limited to five. Experience has confirmed that when there are more than five members in a subgroup, some members become overwhelmed and tend not to participate fully. Finally, before each subgroup's work starts, the members are required to select a moderator and a secretary. The former then presents the subgroup's findings during the final group meeting.

11.3 Data Cleaning and Consolidation

As a critical first step, following the data collection and prior to the data analysis, the raw quantitative data from the questionnaires (or other data collection instruments) must be processed and consolidated in order to be usable. This requires some form of data cleaning, organizing, and coding so that the data is ready to be entered into a database or spreadsheet, analyzed and compared.

Quantitative data is usually collected using a data collection instrument such as a questionnaire. The number of questionnaires or cases is usually fairly large, especially, where probability sampling strategies are used. Due to the nature of quantitative inquiry, most of the questions are closed ended and solicit short 'responses' from respondents that are easy to process and code. It is almost always necessary to use computer software to analyze the data due to a relatively large sample size (number of cases) when compared, for example, to qualitative surveys. Microsoft Excel and Access provide basic spreadsheet and database functions, whereas more specialized statistical software such as SPSS and Epi lnfo can be used where available and where the expertise exists.

Ideally, consolidation and processing is conducted by the same team of interviewers, who completed the data collection (WFP or implementing partner staff or consultants). However, in many cases, additional staff are specifically tasked with the work of entering data into pre-formatted spreadsheets or databases. Data processing and consolidation needs to be closely supervised as careless data entry can significantly affect the quality of subsequent analyses.

Steps to Follow to Consolidate and Process Quantitative Data

The following six steps outline the main tasks related to consolidating and processing quantitative data, prior to analysis:

Step 1: Nominate a Person and Set Procedures to Ensure the Quality of Data Entry

When entering quantitative data into the database or spreadsheet, set up a quality check procedure such as having someone who is not entering data check every 10th case to make sure it has been entered correctly.

Step 2: Enter Numeric Variables on Spreadsheets

Numeric variables should be entered into the spreadsheet or database with each variable on the questionnaire making up a column and each case or question making up a row. The type of 'case' will depend on the unit of study (e.g. individual, households, school, or other).

Step 3: Enter Continuous Variable Data on Spreadsheets

Enter raw numeric values for continuous variables (e.g., age, weight, height, anthropometric scores, income). A new categorical variable can be created from the continuous variable later to assist in analysis. For two or more variables that will be combined to make a third variable, one must be sure to enter each separately. The intent is to ensure that the detail is not lost during the data entry so that categories and variable calculations can be adjusted later, if need be.

Step 4: Coding and Labeling Variables

One must code the categorical nominal variables numerically (e.g., give each option in the variable a number). Where the variable is ordinal (e.g., defining a variable's position in a series), one must be sure to order the codes in a logical sequence (e.g., 1 equals lowest and 5 equals the highest). In SPSS and some other software applications, it is possible to give each numeric variable a value label (e.g., the nominal label that corresponds with the numeric code). For Excel and other software that do not have this function, create a key for each nominal variable that lists the numeric codes and the corresponding nominal label.

Step 5: Dealing with a Missing Value

One must be sure to enter an improbable number, such as 9 for cases in which the probable answers range from 1 to 5, 6, 7 etc. The cell must not be left blank as a blank cell indicates a missing value (e.g. the respondent did not answer the question; the interviewer skipped the question by mistake; the question was not applicable to the respondent; or the answer was illegible). It is best practice to code missing values as 99, 999, or 9999. One must also make sure the number of

9's used makes the value an impossible value for the variable (e.g. for a variable that is 'number of cattle', use 999 since 99 cattle may be a plausible number in some areas). It is important to code missing values so that they can be excluded during analysis, on a case by case basis. For example, by setting the missing value outside the range of plausible values, one can selectively exclude it from analysis in any of the computer software packages (described above).

Step 6: Data Cleaning Methods

Even with quality controls, it will be necessary to 'clean the data', especially, for large data sets with many variables and cases. This allows for obvious errors in data entry to be corrected as well as for the responses to be excluded that simply do not make sense. (Note that the majority of these should be caught in data collection, but even the best quality control procedures miss some mistakes.) To clean the raw data, run simple tests on each variable in the dataset. For example, a variable denoting the sex or gender of the respondent (1 = male, 2 = female) should only take values 1 or 2. If a value such as 3 shows up, a data entry mistake has occurred. Also one must look for impossible values (outside the range of plausibility) such as a child weighing 100 kg, a mother being 10 years old, or a mother being a male, etc.

An example follows of quantitative data consolidation and processing through the application of the six steps outlined above. In this example, each household is the unit of study for the survey and is considered a case.

Step 1: Nominate a person and define a procedure to ensure the quality of data entry. For example, every 4th case will be checked by a non-data entry person (example: field editor) to ensure quality in data entry.

Step 2: Enter numeric variables on spreadsheets and

Step 3: Enter continuous variable data on spreadsheets

Q1: The estimated expenditure on food in the last 6 months responses on questionnaires:

Case 1 $30
Case 2 $23
Case 3 $112
Case 4 $40

Q2: The estimated total expenditure in the last 3 months responses on questionnaires:

Case 1 $50
Case 2 $35
Case 3 $140
Case 4 $35

Enter into the database or spreadsheet and create (derive) a third categorical variable that is food expenditure as a percentage of total expenditure.

Table 7: Expenditure

	Food Expenditure in Last 3	Total Expenditure In Last in Last 3	Food Exp. as a % of total
Case 1	$30.00	$50.00	60%
Case 2	$23.00	$35.00	65.7%
Case 3	$112.00	$140.00	80%
Case 4	$40.00	$35.00	114.3%

Step 4: Coding and Labeling Variables

Coding is the process by which the information that has been obtained from questionnaires, or any other investigation, is translated into a form that can be analyzed by the use of a computer program. Coding involves the assigning of a specific value to the given information in a questionnaire, which is then usually given a label. Coding can also make the data more consistent. For instance, if a question "marital status" is asked, the responses might be "married", "divorced", "separated", "single", "widowed", or "M", "D", "S", "S", "W", among others. Numeric coding will help avoid misinterpretation.

Coding systems

A coding system (code and label) which is common for dichotomous variables is the following:

    0 = No   1 = Yes,

where the number 1 stands for the value assigned, and Yes stands for the label or the meaning of that value. Other users prefer using the system of 1 and 2, that is

    1 = No   2 = Yes.

This means that in coding, it is very important to make the meaning of the value assigned to a piece of information very clear. It can be seen in the two cases above that the ones used have different meanings, that is, in the first case, 1= Yes, whereas in the second case 1 = No. Both are correctly used and are each dependent on how the data was coded. A user may make this clear, by having a data dictionary created as a separate file accompanying the data set.

Coding: Dummy Variables

Dichotomous variables can sometimes also be dummy variables. A "dummy" variable is any variable coded to have two levels, such as the yes/no variables as well as male/female variables. These can also be used to stand in for or to represent, more complicated variables. This is very useful, especially, when a researcher has many values that are much more meaningful when analyzed in terms of response involving yes/no.

For example, a researcher may have collected data on the number of cigarettes that are smoked per week, with sixty five (65) different responses, which range from no cigarettes smoked at all to 3 packs smoked a week, but these data can be decoded as a dummy variable as follows: 1 = Smokes (at all), 0 = Non-smoker. This can also be applied to education: 0 = No post-high school education, 1 = any post-high school education; food consumption: 0 = did not eat the item, 1 = Ate item during time period, and many other variables. Such coding is very useful during the later stages of analysis.

Attaching Labels to Values

Most analysis software packages enable a user to attach a label to variable values. The computer then labels the 0's as male and the 1's as female automatically, which makes it much easier when looking at the output, as shown in the example below:

	Variable SEX	Frequency	Percent
Without label:	0	21	60%
	1	14	40%
With label:	Variable SEX	Frequency	Percent
	Male	21	60%
	Female	14	40%

Coding of Ordinal Variables

The coding process of ordinal variables is similar to the other categorical variables. For example, the variable EDUCATION mentioned above might be coded as follows:

0 = did not graduate from high school
1 = High school graduate
2 = Some college or post-high school education
3 = College graduate.

It should be noted that there should be consistency in numbering for this ordinal categorical variable because the value of the code that is assigned has significance. The higher code represents the more educated the respondent and vice versa. The coding of the variable EDUCATION can also alternately be done in reverse, so that 0 = college graduate, 1 = some college or post-high school education, and so on. This would imply that the higher code represents the less educated respondent. Both the first and the reverse cording are just fine, so long as the coding used is taken into consideration when interpreting the analysis.

The following represents an example of bad coding:

0 = some college or post-high school education
1 = High school graduate
2 = College graduate
3 = did not graduate from high school

The data that is being coded has an inherent order, but the coding given in this example does not seem to follow that order. Such a coding is *not* appropriate for an ordinal categorical variable.

Coding of Nominal Variables

The order is not very important and makes no difference in a nominal categorical variable. Although each category is coded with a number, the number chosen does not represent a numerical value. For example, a variable RESIDE could be coded as follows:

1 = Northeast
2 = South
3 = Northwest
4 = Midwest
5 = Southwest.

The order that is used for these categories does not matter, as opposed to the ordinal category. Northwest can be coded as 1, 2, 4 or 5 because there is no ordered value that is associated with each response.

Coding of Continuous Variables

In continuous variables, the coding is straightforward. Suppose someone gives his age as 48 years, it is entered into the database as 48. However, what if a researcher decides to use age *categories* rather than the data collected and expressed in actual years?

It is a common practice to create categories from a continuous variable by the use of analysis software. With a software package, a researcher can break down a continuous variable into categories by first creating an ordinal categorical variable, as shown below:

AGE CAT
1 = 0–9 years old
2 = 10–19 years old
3 = 20–39 years old
4 = 40–59 years old
5 = 60 years or older

A researcher may also find it easier to code responses from open-ended and fill-in-the-blank questions. For example, in an open-ended question such as "Why did you decide not to take the money to the bank?" sometimes it may result in different answers from respondents. In some cases, the researcher may provide choices to the responses for a particular question as well as an "other (specify)" option, whereby a respondent has the freedom to write any response of his/her choice. Such open-ended questions create more work during the analysis. Therefore, it is always advisable to group responses together that have similar themes when analyzing the information. For instance, the open-ended question above can be answered as *"Didn't think the money was worth taking to the bank," "Was not expecting more money,"* or *"Was planning to use the money immediately"* – these could all be grouped together. It is also necessary to code responses such as *"Don't know".*

It is also very important to note that, although, coding is usually not done until the data has been gathered; a researcher should think how the coding will be done when designing the questionnaire, that is, before the data is gathered. This helps the researcher to be able to collect the data in a user friendly format.

Q3: Name of Village (with corresponding numeric code added)

Case 1 Hagadera = 1
Case 2 Hagadera = 1
Case 3 Kulan = 2
Case 4 Bardera = 3

Q4: Highest level of education completed by the head of household (with corresponding ordinal numeric codes that reflect least education to most)

Case 1 some primary, did not complete = 2
Case 2 no formal schooling = 1
Case 3 completed primary, some secondary = 4 Case 4 completed primary 3

Enter into database: *Table 8: Education Levels*

	Food Expenditure in thr Last 3 Months	Total Expenditure in Last 3 Months	Food Exp as a % of total Expenditure	Village	Highest ed. Level Completion by HofHH
Case 1	$30.00	$50.00	60.00%	1	2
Case 2	$23.00	$35.00	65.71%	1	1
Case 3	$112.00	$140.00	80.00%	2	4

| Case 4 | $40.00 | $35.00 | 114.29% | 3 | 3 |

Step 5: Missing Values

Q5: Number of children under 5 in household

Case 1 = 2
Case 2 = 0
Case 3 = no answer given (missing value)
Case 4 = 3

Enter into database, giving missing value a value of 99 (we use 99 because, with multiple wives, 9 children under 5 within a household is a possibility, even though it is a remote one for this area).

Table 9: household

	Food Expenditure In Last 3 Months	Total expenditure in Last 3 Months	Food Exp as a % of Total Exp.	Village	Highest ed. Level completion by HofHH	Number of children US in HH
Case 1	$30	$50.	60.00%	1	2	2
Case 2	$23	$35	65.71%	1	1	0
Case 3	$112	$140	8Q.00%	2	4	99*
Case 4	$40	$35	114.29%	3	3	3
						*=Not Available

Step 6: Data Cleaning Methods

Before starting the data analysis, it is always necessary to look at (visually scan) the data for any obvious errors, which can result from incorrect data entry. If there is any unexpected very high or very low value (outlier), its validity should be checked. For instance, if the age of teenagers is being investigated; then an age of 36 years or that of 5 years could be an error and should be "cleaned". Suppose in another study 0 = male and 1 = female and an entry was found to have number two. This is clearly an error which should be corrected. Suppose there is a missing value, is it right to assume that the interviewee did not give an answer, or was the missing value not entered into the database (accidentally or otherwise)?

A number of statistical software programs enable the user to set well defined limits during the process of entering data. These can help a user avoid entering outliers such as 2, if the only accepted values are 0, 1, or missing. In some cases, limits may also be set for nominal and continuous variables, such as limiting the digits for age to 3 digits, or limiting the number of words that can be entered.

Procedure

One must run data validity checks to 'clean the data'. Then one must find impossible values for each variable. If found and a review of the source questionnaire does not clarify the mistake, then one must set the value to missing (step 5).

In this case, the third variable in case 4 (refer to the Table under step 5) suggests either an entry error or a mistake on the questionnaire. Food cannot be 114% of total expenditure as food is a portion of expenditure thus the maximum value it could take is 100% (food expenditure represents all expenditure). After reverting to the questionnaire, it was confirmed that the data was entered correctly and that the error lay in the respondent's understanding of the question or in the interviewer's recording of the response. It was decided that the best course of action was to set variables 1, 2, and 3 for Case 4 to 'missing' so that the analysis is not misleading.

Table 10: Expenditure

	Food Expenditure in Last 3 Months	Total Expenditure in Last 3 Months	Food Exp as a % of total Exp.	Village	Highest ed. Level Completion by HofHH	Number of children In HH
Case 1	$30.00	$50.00	60.00	1	2	2
Case 2	$23.00	$35.00	65.71	1	1	0

Case 3	$112.00	$140.00	80.00	2	4	99*
Case 4	9999	9999	999	3	3	3
						*=Not Available

11. 4 Data Analysis

Data analysis is categorized as qualitative and quantitative as discussed below:

Quantitative

- **Univariate**

    In a univariate data analysis, each variable in a data set is explored separately. This is one of the best methods of checking the quality of data. In case, inconsistencies or unexpected results occur, the original data should be used as the reference point when carrying out investigations

- **Multivariate**

    Any statistical technique that is used to analyze data associated with more than a single variable is referred to as a multivariate data analysis. This type of analysis is usually employed whenever the available information can be stored in tables, which contain both rows and columns.

Qualitative

- **Structured questionnaire**

    A structured questionnaire is made up of a series of questions asked to a number of respondents in order to obtain information (about a given topic), which is statistically desirable.

- **Delphi**

    The Delphi technique of data analysis is a forecasting method based on the results of questionnaires sent to a diverse set of experts.

    Other important aspects of data analysis include:

    i. System analysis
    ii. Path analysis
    iii. Log frame analysis
    iv. Quasi experimental analysis

Data Analysis Procedures (DAP)

Data analysis procedures enable a researcher to convert data into understandable information and knowledge as well as to investigate the relationship between variables. Data analysis procedures include:

- Indicator definition
- Relevant strategy identification
- Data transformation
- Direct data interpretation

Data analysis procedures are important because they help in:

i. An appreciation of the scientific method used, meaning hypotheses testing as well as the statistical significance with respect to the research question.

ii. The realization of the importance of a good research design when a research question is being investigated.

iii. Gaining knowledge of a wide range of inferential statistics as well as their applicability and their limitations as far as the research is concerned

iv. Providing the ability to devise, implement as well as to accurately report a small quantitative research study.

v. Allowing the capability to identify the data analysis procedures, which are relevant to the research project

Below is an example of data analysis:

Descriptive analysis of table

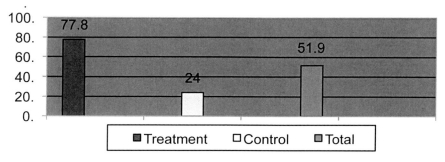

Figure 12

The graph (figure 11) is the output of a survey conducted in one African country. It was part of an evaluation exercise, where health service providers were asked if they used graphs in their monitoring process. The graph table was generated using EPIINFO software.

There were two groups of health service providers (HSPs) from different health facilities spread across the country. These facilities were classified into two groups: intervention (treatment) and non-intervention (control). The facilities were randomly (for convenience) selected. In the former, HSPs were trained on logistics management of contraceptives, while in the latter group, service providers continued to use experience gained from other (earlier) programs. This survey was conducted about one year after the formal training in the treatment arm. This timeline was established in order to give trainees an opportunity to adequately use the skills that they had acquired during the training. During this interval, the regular supervision by the ministry of health senior staff was performed in both arms as usual.

During the analysis, different statistical tests were performed. One of them is presented in Table 9. The test performed here was the chi-square test. This test is based on frequency distribution and, generally, used to test for differences between two groups (in this case treatment and control). The test is also complemented by a hypothesis (in this case, "the two groups are statistically different"). The hypothesis is then either rejected or not rejected.

Table 11: Health service providers

Graphs for monitoring	1	2	Total
1 Row% Col%	21 77.8 77.8	6 22.2 24.0	27 100.0 51.9
2 Row% Col%	6 24.0 22.2	19 76.0 76.0	25 100.0 48.1

Total	27	25	52
Row%	51.9	48.1	100.0
Col%	100.0	100.0	100.0

(1= TRAINED IN LOGISTICS MANAGEMENT AND 2= NOT TRAINED in LOGISTICS MANAGEMENT)

2. The respondents were asked if they used graphs for monitoring

3. In the columns, 1= treatment group (those trained in Logistics Management (LM) and 2 = Control Group (not trained in LM)

4. The rows 1= yes (meaning use graphs for monitoring) and 2=No (Do not produce graphs during monitoring.

The results of the graph are as follows:

A total of 52 health facilities (HFs) were included in the survey. 27 were from the treatment circuit while 25 belonged to the control arm. Over half of the respondents (51.9%) answered "yes" to the question. In this group, 77.7% belonged to the treatment arm with the rest coming from the control group. The chi-square ($p < .0001$) test was also used. The conclusion of the analysis was that in HFs where HSPs were trained in contraceptive logistics management, staff were more likely to use graphs as part of their monitoring than they were in the control group, where a similar training exercise was not conducted. There is, therefore, insufficient evidence to reject the hypothesis that the two groups were different. What this implies is that the training program was effective and that a similar exercise performed in other HFs might well improve monitoring efforts in HFs.

Case Study: Qualitative Analysis – Focus group Discussion (FGD)

The following section presents details of an evaluation exercise that I conducted. The sessions were all moderated and presented in french.

The health facility where the intervention (refurbishing and capacity development) took place was jointly supported by a major multilateral agency (funding) and the government of Vanuatu (implementing partner), a Pacific island country (PIC).

The project outcome to which this intervention was contributing was: "the increased utilization of high-quality, comprehensive sexual and reproductive health information and services, including comprehensive HIV prevention services, particularly for most at risk populations, including young people".

The Port Orly (a district on the island of Santo province in Vanuatu) Health Center (HC) provides comprehensive services ranging from an outpatient department to a maternity room. Two staff members, a midwife (trained as part of this project) and an advanced nursing practitioner, manage the center. Services are provided seven days a week with 24-hour emergency services also available. The center's coverage area includes between 50 and 60 communities and about 7,000 people. A major problem with the center was the lack of an appropriate client waiting room: an issue that the project was supposed to address.

Due to time limitation, I organized two focus group discussion sessions that lasted two and a half hours with the first group and three and a half hours with the second. The target audience of the first group was the two service providers (supply) identified above. The second group was the beneficiaries (demand): women of reproductive age and specifically pre-natal (new and repeat) clients. The selection of participants in the second group was done by staff before I arrived. They were all in attendance as planned.

The exercise with the first group was limited to the two service providers and no one else. During this session, we examined all the registers and calculated summaries from the previous year's routine statistics. I then moved on to the FGD exercise starting with the code of ethics and informed them that the entire exercise was anonymous. I stressed the fact that no names would be used without participant authorization.

In summary, during the session, the staff expressed their respective appreciation to the donor for providing this training to the staff and sprucing up their HF. The picture of the outcome is presented in figure 11a. Both consistently agreed that the project did not only increase the client load (as confirmed by the service statistics), but it also raised morale, motivation and gave them more of a sense of pride when executing their respective daily tasks. Since the completion of the project about a year earlier, the center had experienced an increase in its client load by 30 percent and an increase in the daily outpatient load of over 50 percent.

Specific questions asked were unprompted using a modified questionnaire provided by DAC. The questionnaire is basically divided into different sections (sustainability, impact, relevance, effectiveness, efficiency) of the evaluation criteria. In all the responses, both members were unanimous in their responses: positive and complimentary. They also indicated that, for most cases, they were quite satisfied, though other areas of concern existed. For example, they highlighted the inadequate and timely delivery of commodities, including some that had already expired. The other area of concern (which was also confirmed by the beneficiaries) was the limited number of client chairs that had been bought as part of the renovation process. Twenty more chairs were required to accommodate most of their clients. Only ten seats were supplied as part of the intervention. This was noted in the report and the relevant parties promised to fill in that gap. They further indicated that, as a result of the shortage and even before the project, clients had a tendency to roam around until they were able to receive the services they needed. In many cases, some beneficiaries got tired or frustrated, went home, and never returned. This is no longer an issue.

The second FCD excluded the service providers (SPs) for obvious reasons. The methodology was similar except for the exclusion of the DAC questionnaire as it was considered inappropriate and irrelevant. The initial level of participation in this group was cool and distant, which suddenly changed as the twenty-four participants gradually recognized the importance of the session and started being more enthusiastics.

The formal session started with general discussions that ranged from quality service (QS) to access to medication. With regard to QS, the responses were mixed. The minority of them felt that the professional staff members were heavy handed, rigid and condescending whereas the majority expressed contradictory views. In general, the group applauded the SPs for their dedication, caring and professionalism.

As discussions focused more and more on specifics, the participants' unanimous endorsement and approval was overwhelming in terms of their satisfaction and gratitude towards the funding agency of the project. They unanimously agreed that the rehabilitation was a timely, constructive, and effective intervention. Nonetheless, they did highlight the inadequate number of seats available. Furthermore, they pointed out (as demonstrated in the picture) that many of them were forced to sit on the floor because of the limited number of chairs. They all reminisced about the past when there was no waiting space or chairs; and how they were forced to roam around waiting for their respective turns to be served. Also, consistent with what was expressed by the first group, they highlighted occasions when some had to return home out of frustration. They requested that a supply of 20 additional seats be procured to accommodate the increased number of people in the waiting area. This request was shared with the relevant parties. Based on the discussion with these groups and the increased client load, one can only conclude that the center's rehabilitation will have sustainable effects as the structure is part of an existing facility that is already well managed.

Fig 12a: Port Orly Health Facility

Fig 12b: Cross-Section of Port Orly Health Facility Ante-Natal Clients

11.5 Report Writing

A report is an important management tool, which may influence future actions, policies, and/or procedures. A good report should be reader friendly and relatively brief; otherwise, it may not be read at all. A report should bear the following:

- Key findings
- Lessons learned
- Recommendations
- Way forward - a proposed Plan of Action

11.6 Decision making

Decision making is a process that includes information gathering, analyzing, synthesizing, processing, screening and ultimately using the collected data to accomplish an objective. It is also the process of reviewing, revising and implementing the relevant changes that can enhance the achievement of intended plans.

It is an activity than can either be based on quantitative or qualitative data. In some cases where these types of data are either unreliable or unavailable, experience and consultative processes can also be useful. Some of the factors considered in many decision making processes include:

Institution mandates:
- Relevance of findings
- Planning
- Effects on on-going strategies
- Review and appropriate realignment policies, procedures, etc.

Relevance of Findings:

In any decision making process, one important consideration is its relevance with regard to influencing the final outcome. The strength of the evidence available serves as a significant and crucial element that contributes significantly to any decision making process. It is, therefore, critical and important to initially identify the evidence-anecdotal or scientific - and then assess its usefulness with regard to the intended outcome. Such a strategy helps solidify any decision dynamic and enhances the chances of mitigating potential weaknesses. An effective and appropriate relevance also contributes to strengthening one's potential of achieving one's planned objectives.

Planning:

The success of any initiative depends significantly on the level and degree of establishing an effective plan. Plans serve as a road map to help the planners better structure their vision and, therefore, minimize any chances and risks involved in executing their relevant strategies. Plans are also a framework that if monitored well, contribute to aligning one's objectives to one's intended targets. Sometimes ambiguities exist between plans and strategies. In this context, plans represent the "*what*", while strategies stand for the "*how*". A simplistic and straight forward interpretation like this helps delineate a trade off between these two words. Hence, a good and effective plan is necessary, although, it is not sufficient. A strategy, therefore, serves as a "driving force" that is required to move the activities from baseline to midline to end-line.

Effects on On-Going Strategies:

Strategies from a monitoring and an evaluation point of view are dynamic. They are strongly driven by existing risks and assumptions so they remain potentially subject to change. When conditions and landscapes whether political, environmental, social, or economic manifest themselves over time, one's strategies need to reflect these changes. There is a misconception among program managers and stake holders that strategies should never change and should be implemented from program design to the end of implementation.

Review and Appropriate Realignment of Policies and Procedures:

This process simply confirms and better serves as a close reminder and re-definition of the preceding section. In program implementation and management, different and relevant stakeholders need to constantly monitor their bench marks and revise or refine them to make sure that they are consistent with the revised strategies. This is in order to better manage their activities and achieve meaningful and intended results. Existing policies and guidelines are subject to change. And in some cases, these changes are driven by either "internal" or "external" factors. The former represents changes within program management control; while the latter, are changes driven by outside forces. In both cases, the strategies may vary significantly and hence the importance and existence of an effective oversight cannot be adequately emphasized. Change in strategy is a necessary "evil" in any program management. Hence collectively, all these elements contribute to helping everyone make informed decisions.

Conclusion

This document has outlined the various aspects of a typical M and E and data management systems. The latter included basic statistics, a data management framework and model, and illustrated various applications of several statistical software packages. The principals of monitoring and evaluation were also covered. The M and E section, like data management, has been reinforced with real life examples and case studies. Such an approach is meant to demystify the myths that are sometimes deterrents to applying and correctly understanding the ongoing principles.

This text should at best have helped the lay person and other experts design and manage their projects more effectively and competently. A successfully implemented program is good for all parties – the beneficiaries, stake holders, implementing partners and funding agencies. "Success" should continue to be our complete mantra. I hope this manual has served as a beneficial guide for those who need to implement the types of monitoring and evaluation components to achieve their intended results. I aim for this text to complement any of the now available powerful statistical tools on the market which turn raw data into meaningful information concerning what is actually being accomplished, relative to program objectives.

I hope I have successfully produced a book that caters to a vast range of audiences starting with different skill and experience levels, from the student first trying to gain an appreciation of what "management" involves in health and development programs, to the experienced field practitioner trying to determine specific M & E approaches to apply to his/her current project. There is also provision for the very experienced parties who simply want to have easy access to readily available and useful information on M and E and data management systems.

Finally, I have always remembered one incident when some of my national colleagues who were professionals in their own right, continuously thanked me for showing them what they had often been taught in the class room, but had never had the chance to practice. Sampling and its protocol was a case that came up quite often. It is, therefore, my hope that this primer has effectively addressed these issues and that I have correctly represented what one needs to learn in a constructive, simple and comprehensible way.

Appendix

Appendix 1: Data Management Framework

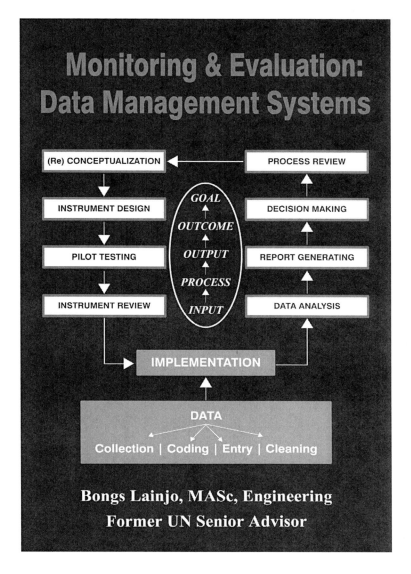

Appendix 2a: Program Indicator Screening Matrix (PRISM)

Programme Indicator Screening Matrix (PRISM)

Thematic Area: RH, PDS, GDR, Other Results Level: Goal, Outcome, Output INDICATOR	1 Specificity	2 Reliability	3 Sensitivity	4 Simplicity	5 Utility	6 Affordablity	7 Total Yes	% Score	Implemented Yes/No

Specificity - Does it measure the result?
Reliability - Is it consistent measure over time?
Sensitivity - When the result changes will it be sensitive to those changes?
Simplicity - Will it be easy to collect and analyze the Data?
Utility - Will the information be useful for decision-making and learning?
Affordability - Can the program/project afford to collect the Data?

Appendix 2b: PRISM case-study – Intra-group Screening

PROGRAM INDICATOR SCREENING MATRIX (PRISM) WORKSHOP INTRA-WORKING GROUP GUIDELINES

GROUP NAME:
GROUP MODERATOR:
GROUP RAPPORTEUR:
NUMBER OF GROUP PARTICIPANTS:
RESULTS LEVEL: GOAL, OUTCOME, OUTPUT
THERMATIC ARE: RH, PDS, GDR

BACKGROUND:

- During the most recent Mid Term Review (MTR), an assessment of the current country programme was made. Key recommendations were made including the realignment of certain interventions and fine tuning of objectives - results.

TASK:

- Each group is given a number of output indicators – level at which UNFPA is accountable (one of its key pillars). If these indicators are well defined and accomplished during programme implementation, they will contribute to either the **outcome** or the **goal**. The task of each group therefore will be to further refine the **output** indicators through a screening process using the PRogramme Indicator Screening Matrix (PRISM) provided. PRISM is a matrix of indicators and selection criteria that if used judiciously can substantially minimize redundant indicators and as a result yield more constructive deliverables and contribute significantly to higher-level results.

Appendix 2c: PRISM case study – Intra-group screening

PROGRAM INDICATOR SCREENING MATRIX (PRISM) INTRA-WORKING GROUP GUIDELINES (continued...)

OUTPUT:

- At the end of each group work, the following results will be presented:
- Number and type of indicators by criterion unanimously recommended by the group;
- Number and type of indicators by criterion with discordant (indicate number in favor and number against) views;
- Number and type of indicators (overall) unanimously recommended for implementation;
- Number and type of indicators (overall) with discordant (indicate number in favor and number against) views;
- A follow up action plan of how the performance of the recommended indicators and lessons learnt will be periodically monitored by each group

Appendix 2d: PRISM case-study – Inter-group Screening

PROGRAM INDICATOR SCREENING MATRIX (PRISM)
INTER-WORKING GROUP GUIDELINES

GROUP NAME:
GROUP MODERATOR:
GROUP RAPPORTEUR:
NUMBER OF GROUP PARTICIPANTS:
RESULTS LEVEL: GOAL, OUTCOME, OUTPUT
THERMATIC ARE: RH, PDS, GDR

BACKGROUND:
- During the most recent Mid Term Review (MTR), an assessment of the current country programme was made. Key recommendations were made including the realignment of certain interventions and fine tuning of objectives - results.

TASK:
- Each group is given a number of output indicators – level at which UNFPA is accountable (one of its key pillars). If these indicators are well defined and accomplished during programme implementation, they will contribute to either the outcome or the goal. The task of each group therefore will be to further refine the output indicators through a screening process using the PRogramme Indicator Screening Matrix (PRISM) provided. PRISM is a matrix of indicators and selection criteria that if used judiciously can substantially minimize redundant indicators and as a result yield more constructive deliverables and contribute significantly to higher-level results.

Appendix 2e: PRISM case study – Inter-group screening (con'd)

PROGRAM INDICATOR SCREENING MATRIX (PRISM)
INTER-WORKING GROUP GUIDELINES (continued...)

OUTPUT:

At the end of each group work, the following results will be presented:

- Number and type of indicators by criterion unanimously recommended by the group;
- Number and type of indicators by criterion with discordant (indicate number in favor and number against) views;
- Number and type of indicators (overall) unanimously recommended for implementation;
- Number and type of indicators (overall) with discordant (indicate number in favor and number against) views;
- A follow up action plan of how the performance of the recommended indicators will be periodically monitored and lessons learned by each group.

Appendix 3

Glossary of Key Terms in M and E

Source: DAC 2010, Glossary of Key Terms in Evaluation, Paris, OECD. Multi language versions (English, French, Chinese, Italian, Japanese, Portuguese, Russian and Spanish) are also available.

Website: http://wbc-inco.net/object/document/7516/attach/Glossaryof KeyTermsinEvaluationandResultsBasedManagement.pdf

Accountability

Obligation to demonstrate that work has been conducted in compliance with agreed rules and standards or to report fairly and accurately on performance results vis a vis mandated roles and/or plans. This may require a careful, even legally defensible, demonstration that the work is consistent with the contract terms.

Note: Accountability in development may refer to the obligations of partners to act according to clearly defined responsibilities, roles and performance expectations, often with respect to the prudent use of resources. For evaluators, it connotes the responsibility to provide accurate, fair and credible monitoring reports and performance assessments. For public sector managers and policy-makers, accountability is to taxpayers/citizens

Activity

Actions taken or work performed through which inputs, such as funds, technical assistance and other types of resources are mobilized to produce specific outputs.

Related term: development intervention.

Analytical tools

Methods used to interpret and process information during evaluation

Appraisal

An overall assessment of the relevance, feasibility and potential sustainability of a development intervention prior to a decision of funding.

Note: In development agencies, banks, etc., the purpose of appraisal is to enable decision-makers to decide whether the activity represents an appropriate use of corporate resources.

Related term: ex-ante evaluation

Attribution

The ascription of a causal link between observed (or expected to be observed) changes and a specific intervention.

Note: Attribution refers to that which is to be credited for the observed changes or results achieved. It represents the extent to which observed development effects can be attributed To a specific intervention or to the performance of one or more partner taking account of other interventions, (anticipated or unanticipated) confounding factors, or external shocks.

Audit

An independent, objective assurance activity designed to add value and improve an organization's operations. It helps an organization accomplish its objectives by bringing a systematic, disciplined approach to assess and improve the effectiveness of risk management, control and governance processes.

Note: a distinction is made between regularity (financial) auditing, which focuses on compliance with applicable statutes and regulations; and performance auditing, which is concerned with relevance, economy, efficiency and effectiveness. Internal auditing provides an assessment of internal controls undertaken by a unit reporting to management while external auditing is conducted by an independent organization.

Base-line study

An analysis describing the situation prior to a development intervention, against which progress can be assessed or comparisons made.

Benchmark

Reference point or standard against which performance or achievements can be assessed.

Note: A benchmark refers to the performance that has been achieved in the recent past by other comparable organizations, or what can be reasonably inferred to have been achieved in the circumstances.

Beneficiaries

The individuals, groups, or organizations, whether targeted or not, that benefit, directly or indirectly, from the development intervention.

Cluster evaluation

An evaluation of a set of activities, project/program activities.

Conclusions

Conclusions point out the factors of success and failure of the evaluated intervention, with special attention paid to the intended and unintended results and strength or weakness. A conclusion draws on data collection and analyses undertaken, through a transparent chain of arguments

Counter factual

The situation or condition which hypothetically may prevail for individuals, organizations or groups were there no development interventions.

Country Program Evaluation/ Country Assistance Evaluation

Evaluation of one or more donor's or agency's portfolio of development interventions, and the assistance strategy behind them, in a partner country.

Data Collection Tools

Methodologies used to identify information sources and collect information during an evaluation.

Note: Examples are informal and formal surveys, direct and participatory observation, community interviews, focus groups, expert opinion, case studies, literature search.

Development Intervention

An instrument for partner (donor and non-donor) support aimed to promote development.

Note: Examples are policy advice, projects, programs.

Development objective

Intended impact contributing to physical, financial, institutional, social, environmental, or other benefits to a society, community, or group of people via one or more development interventions.

Economy

Absence of waste for a given output.

Note: An activity is economical when the costs of the scarce resources used to approximate minimum needed to achieve planned objectives

Effect

Intended or unintended change due directly or indirectly to an intervention.

Related terms: results, outcomes.

Effectiveness

The extent to which the development intervention's objectives were achieved, or are expected to be achieved, taking into account their relative importance.

Note: Also used as an aggregate measure of (or judgment about) the merit or worth of an activity, i.e. the extent to which an intervention has attained, or is expected to attain, its major relevant objectives efficiently in a sustainable fashion and with a positive institutional development impact.

Related term: efficacy.

Efficiency

A measure of how economically resources/inputs (funds, expertise, time, etc.) are converted to results.

Evaluability

Extent to which an activity or a program can be evaluated in a reliable and credible fashion.

Note: Evaluability assessment calls for the early review of a proposed activity in order to ascertain whether its objectives are adequately defined and its results verifiable.

Evaluation

The systematic and objective assessment of an on-going or completed project, programme or policy, its design, implementation and results. The aim is to determine the relevance and fulfillment of objectives, development efficiency, effectiveness, impact and sustainability.

An evaluation should provide information that is credible and useful, enabling the incorporation of lessons learned into the decision–making process of both recipients and donors.

Evaluation also refers to the process of determining the worth or significance of an activity, policy or program. An assessment, as systematic and objective as possible, of a planned, on-going, or completed development intervention.

Note: Evaluation in some instances involves the definition of appropriate standards, the examination of performance against those standards, an assessment of actual and expected results and the identification of relevant lessons.

Related term: Review.

Ex-ante evaluation

An evaluation that is performed before implementation of a development intervention.

Related terms: appraisal, quality at entry.

Ex-post evaluation

Evaluation of a development intervention after it has been completed.

Note: It may be undertaken directly after or long after completion. The intention is to identify the factors of success or failure, to assess the sustainability of results and impacts, and to draw conclusions that may inform other interventions

External evaluation

The evaluation of a development intervention conducted by entities and/or individuals outside the donor and implementing organizations.

Feedback

The transmission of findings generated through the evaluation process to parties for whom it is relevant and useful so as to facilitate learning. This may involve the collection and dissemination of findings, conclusions, recommendations and lessons from.

Findings

Finding uses evidence from one or more evaluations to allow for a factual statement.

Formative evaluation

Evaluation intended to improve performance, most often conducted during the implementation phase of projects or programs.

Note: Formative evaluations may also be conducted for other reasons such as compliance, legal requirements or as part of a larger evaluation initiative.

Related term: process evaluation.

Goal

The higher-order objective to which a development intervention is intended to contribute.

Related term: development objective.

Impacts

Positive and negative, primary and secondary long-term effects produced by a development intervention, directly or indirectly, intended or unintended.

Independent evaluation

An evaluation carried out by entities and persons free of the control of those responsible for the design and implementation of the development intervention.

Note: The credibility of an evaluation depends in part on how independently it has been carried out. Independence implies freedom from political influence and organizational pressure. It is characterized by full access to information and by full autonomy in carrying out investigations and reporting findings.

Indicator

Quantitative or qualitative factor or variable that provides a simple and reliable means to measure achievement, to reflect the changes connected to an intervention, or to help assess the performance of a development actor.

Inputs

The financial, human, and material resources used for the development intervention.

Institutional Development Impact

The extent to which an intervention improves or weakens the ability of a country or region to make more efficient, equitable, and sustainable use of its human, financial, and natural resources, for example through: (a) better definition, stability, transparency, enforceability and predictability of institutional arrangements and/or

(b) better alignment of the mission and capacity of an organization with its mandate, which derives from these institutional arrangements. Such impacts can include intended and unintended effects of an action.

Internal evaluation

Evaluation of a development intervention conducted by a unit and/or individuals reporting to the management of the donor, partner, or implementing organization.

Related term: Self evaluation.

Joint evaluation

An evaluation to which different donor agencies and/or partners participate.

Note: There are various degrees of "jointness" depending on the extent to which individual partners cooperate in the evaluation process, merge their evaluation resources and combine their evaluation reporting. Joint evaluations can help overcome attribution problems in assessing the effectiveness of programs and strategies, the complementarities of efforts supported by different partners, the quality of aid coordination etc.

Lessons learned

Generalizations based on evaluation experiences with projects, programs, or policies that abstract from the specific circumstances to broader situations. Frequently, lessons highlight strengths or weaknesses in preparation, design, and implementation that affect performance, outcome, and impact.

Logical framework (Logframe)

Management tool used to improve the design of interventions, most often at the project level. It involves identifying strategic elements (inputs, outputs, outcomes, impact) and their causal relationships, indicators, and the assumptions or risks that may influence success and failure. It thus facilitates planning, execution and evaluation of a development intervention.

Related term: results based management.

Meta-evaluation

The term is used for evaluations designed to aggregate findings from a series of evaluations. It can also be used to denote the evaluation of an evaluation to judge its quality and/or assess the performance of the evaluators.

Mid-term evaluation

Evaluation performed towards the middle of the period of implementation of the intervention.

Related term: formative evaluation.

Monitoring

A continuing function that uses systematic collection of data on specified indicators to provide management and the main stakeholders of an ongoing intervention with indications of the extent of progress and achievement of objectives and progress in the use of allocated funds.

Related term: performance monitoring, indicator.

Outcome

The likely or achieved short-term and medium-term effects of an intervention's outputs.

Related terms: result, outputs, impacts, effect.

Outputs

The products, capital goods and services which result from a development intervention; may also include changes resulting from the intervention which are relevant to the achievement of outcomes.

Participatory evaluation

Evaluation method in which representatives of agencies and stakeholders (including beneficiaries) work together in designing, carrying out and interpreting an evaluation.

Partners

The individuals and/or organizations that collaborate to achieve mutually agreed upon objectives.

Note: The concept of partnership connotes shared goals, common responsibility for outcomes, distinct accountabilities and reciprocal obligations. Partners may include governments, civil society, non-governmental organizations, universities, professional and business associations, multilateral organizations, private companies, etc.

Performance

The degree to which a development intervention or a development partner operates according to specific criteria/standards/guidelines or achieves results in accordance with stated goals or plans.

Performance indicator

A variable that allows the verification of changes in the development intervention or shows results relative to what was planned.

Related terms: performance monitoring, performance measurement.

Performance measurement

A system for assessing performance of development interventions against stated goals.

Related terms: performance monitoring, indicator.

Performance monitoring

A continuous process of collecting and analyzing data to compare how well a project, program, or policy is being implemented against expected results.

Process evaluation

An evaluation of the internal dynamics of implementing organizations, their policy instruments, their service delivery mechanisms, their management practices, and the linkages among these.

Related term: formative evaluation.

Program evaluation

Evaluation of a set of interventions, marshaled to attain specific global, regional, country, or sector development objectives.

Note: a development program is a time bound intervention involving multiple activities that may cut across sectors, themes and/or geographic areas.

Related term: Country program/strategy evaluation.

Project evaluation

Evaluation of an individual development intervention designed to achieve specific objectives within specified resources and implementation schedules, often within the framework of a broader program

Note: Cost benefit analysis is a major instrument of project evaluation for projects with measurable benefits. When benefits cannot be quantified, cost effectiveness is a suitable approach.

Project or program objective

The intended physical, financial, institutional, social, environmental, or other development results to which a project or program is expected to contribute.

Purpose

The publicly stated objectives of the development program or project.

Quality Assurance

Quality assurance encompasses any activity that is concerned with assessing and improving the merit or the worth of a development intervention or its compliance with given standards.

Note: examples of quality assurance activities include appraisal, RBM, reviews during implementation, evaluations, etc. Quality assurance may also refer to the assessment of the quality of a portfolio and its development effectiveness.

et un plan de travail déterminés, souvent dans le cadre d'un programme plus large.

Reach

The beneficiaries and other stakeholders of a development intervention.

Related term: beneficiaries.

Recommendations

Proposals aimed at enhancing the effectiveness, quality, or efficiency of a development intervention; at redesigning the objectives; and/or at the reallocation of resources. Recommendations should be linked to conclusions.

Relevance

The extent to which the objectives of a development intervention are consistent with beneficiaries' requirements, country needs, global priorities and partners' and donors' policies.

Note: Retrospectively, the question of relevance often becomes a question as to whether the objectives of an intervention or its design are still appropriate given changed circumstances.

Reliability

Consistency or dependability of data and evaluation judgements, with reference to the quality of the instruments, procedures and analyses used to collect and interpret evaluation data.

Note: evaluation information is reliable when repeated observations using similar instruments under similar conditions produce similar results.

Results

The output, outcome or impact (intended or unintended, positive and/or negative) of a development intervention.

Related terms : outcome, effect, impacts.

Results Chain

The causal sequence for a development intervention that stipulates the necessary sequence to achieve desired objectives-beginning with inputs, moving through activities and outputs, and culminating in outcomes, impacts, and feedback. In some agencies, reach is part of the results chain.

Related terms: assumptions, results framework.

Results framework

The program logic that explains how the development objective is to be achieved, including causal relationships and underlying assumptions.

Results-Based Management (RBM)

A management strategy focusing on performance and achievement of outputs, outcomes and impacts.

Related term: logical framework.

Review

An assessment of the performance of an intervention, periodically or on an ad hoc basis.

Note: Frequently "evaluation" is used for a more comprehensive and/or more in-depth assessment than "review". Reviews tend to emphasize operational aspects. Sometimes the terms "review" and "evaluation" are used as synonyms.

Related term: evaluation.

Risk analysis

An analysis or an assessment of factors (called assumptions in the logframe) affect or are likely to affect the successful achievement of an intervention's objectives. A detailed examination of the potential unwanted and negative consequences to human life, health, property, or the environment posed by development interventions; a systematic process to provide information regarding such undesirable consequences; the process of quantification of the probabilities and expected impacts for identified risks.

Sector program evaluation

Evaluation of a cluster of development interventions in a sector within one country or across countries, all of which contribute to the achievement of a specific development goal.

Note: a sector includes development activities commonly grouped together for the purpose of public action such as health, education, agriculture, transport etc.

Self-evaluation

An evaluation by those who are entrusted with the design and delivery of a development intervention.

Stakeholders

Agencies, organisations, groups or individuals who have a direct or indirect interest in the development intervention or its evaluation.

Summative evaluation

A study conducted at the end of an intervention (or a phase of that intervention) to determine the extent to which anticipated outcomes were produced. Summative evaluation is intended to provide information about the worth of a program.

Related term: impact evaluation.

Sustainability

The continuation of benefits from a development intervention after major development assistance has been completed.

The probability of continued long-term benefits. The resilience to risk of the net benefit flows over time.

Target group

The specific individuals or organizations for whose benefit the development intervention is undertaken.

Terms of reference

Written document presenting the purpose and scope of the evaluation, the methods to be used, the standard against which performance is to be assessed or analyses are to be conducted, the resources and time allocated, and reporting requirements. Two other expressions sometimes used with the same meaning are "scope of work" and "evaluation mandate".

Thematic evaluation

Evaluation of a selection of development interventions, all of which address a specific development priority that cuts across countries, regions, and sectors.

Triangulation

The use of three or more theories, sources or types of information, or types of analysis to verify and substantiate an assessment.

Note: by combining multiple data-sources, methods, analyses or theories, evaluators seek to overcome the bias that comes from single informants, single-methods, single observer or single theory studies.

Validity

The extent to which the data collection strategies and instruments measure what they purport to measure.

BIBLIOGRAPHY

Agresti, A. & Finlay, B. 1997. *Statistical Methods for the Social Sciences*, 3th Edition.

Alston, M. and Bowles, W. 1998. *An Introduction to Research for social workers*

Anderson, T. and Sclove, L. 1974. *Introductory Statistical Analysis.* Houghton Mifflin

BoS. *Basics of Statistics.*www.mv.helsinki.fi/home/jmisotal/BoS.pdf

Case studies. http://cscar.research.umich.edu/case-studies

Clarke, G. & Cooke, D. 1998. A *Basic course in Statistics.* Arnold Company.

de Mendoza, A. 2010. *Monitoring and Evaluation Framework* (2010-2013): Fund for Gender Equality

Diakonia (no date). *Planned Advocacy Assessment Tool: a Tool for Assessing Your Organization's Strengths and Weaknesses in Planning Advocacy Activities.*

Farmer, R., Miller, D. and Lawrenson, R., 1996. *Epidemiology and Public Health*

Freeman.

Freund, J.E., *Modern Elementary Statistics.* Prentice-Hall, 2001.

Health and social care practitioners. 2nd Edition. London : Continuum

Harvard University, EDX MOOC, 2012

Hek, G., Judd, M. and Moule, P. 2002. *Making Sense of Research. An introduction for*

http://betterevaluation.org/evaluation-options/dac_criteria. Accessed on 11th June 2014.

http://highered.mcgraw-hill.com/sites/dl/free/0073373605/582605/Chapter16_SamplingMethods.pdf

http://www.learntech,uwe.ac.uk/da/Default/aspx.

http://www.mv.helsinki.fi/home/jmisotal/BoS.pdf

http://www.stats.gla.ac.uk/steps/glossary/confidence_intervals.html
http://www.stats.gla.ac.uk/steps/glossary/sampling.html - Accessed 20/01/09
http://www.statsoft.com/textbook/stathome.html.
https://www.csulb.edu/~msaintg/ppa696/696stsig.htm
IFAD (2002): *Managing for Impact in Rural Development: A Guide for Project M&E.*
INTER AMERICAN DEVELOPMENT BANK (IADB): A Management Tool for Improving Project Performance. <Online> Available at: www.iadb.org/cont/evo/EngBook/eng book.htm
Jeffrey, L. 2013. *Data Scientist Insights: Six Types Of Analyses Every Data Scientist*
Johnson, A. and Bhattacharyya, G. 1992. *Statistics: Principles and Methods*, 2nd Edition.
Johnson, R. and Bhattacharyya, K. 1992. *Statistics: Principles and Methods, 2nd Edition.*
Kaplan, J. 2014. *Evaluation options: better evaluation.* <Online> Available at
Lainjo, B. 2013. *Program Indicator Screening Matrix* (PRISM): *A Composite Score Framework.* Canadian Evaluation Society (CES) Conference Toronto, Canada
Leppälä, R. 2000. Ohjeita tilastollisen tutkimuksen toteuttamiseksi SPSS for Windows –
Medicine (4th ed.), Blackwell Science, Oxford.
methods. Australia: Allen and Unwin
Moore, D. 1997. *The Basic Practice of Statistics.* Freeman.
Moore, D. and McCabe, G. 1998. *Introduction to the Practice of Statistics, 3th Edition.*
Newbold, P. 1995. *Statistics for Business and Econometrics.* Prentice Hall.
OECD/DAC (1991): *Principles for Evaluation of Development Assistance.* Available at: www.oecd.org/dac/Evaluation/pdf/ evalprin.pdf

OECD/DAC (1998): *Effective Practices in Conducting a Joint Multi-Donor Evaluation.* Available at: www.oecd.org/dac/

OECD/DAC (1998): *Review of the DAC Principles for Evaluation of Development Assistance.* Available at: www.oecd.org/dac/ Evaluation/pdf/eval.pdf

OECD/DAC: *Evaluation Criteria.* Available at: www.oecd.org//dac/ Evaluation/htm/evalcrit.htm

OECD/PUBLIC MANAGEMENT SERVICE (1999*): Improving Evaluation Practices: Best Practice Guidelines for Evaluation and Background Paper.* Available at: www.oecd.org/puma

OECD/WORKING PARTY ON AID EVALUATION (2001): Glossary of Terms in *Evaluation and Results-Based Management.* Available at: www.oecd.org/dac/htm/glossary.htm

ohjelmiston avulla, Tampereen yliopisto, Matematiikan, tilastotieteen ja filosofian laitos, B53.

OECD (f). *Aid Effectiveness: Three Good Reasons Why the Paris Declaration Will Make a*

Difference. 2005 Development Cooperation Report. Vol. 7. No. 1 2006 49-54.

OECD (c). 2005 *Development* Cooperation Report Vol. 7 No. 1 p. 54. 2006

OECD (a). *Aid Effectiveness2005-2010: Progress in Implementing the Paris Declaration.* OECD

Publishing. 2011
<http://www.oecd.org/development/aideffectiveness/48742718.pdf> .

OECD (d). *Paris Declaration Indicators of Progress, in OECD, Aid Effectiveness 2011:*

Progress in Implementing the Paris Declaration, OECD Publishing. 2012;

http://www.oecd.org/development/aideffectiveness/48742718.pdf

OLIVE (2002): *Planning for Monitoring and Evaluation*. The Monitoring and Evaluation Handbook.

Organization for Economic Cooperation and Development (OECD). 1998. *"Review of the DAC Principles of Development Assistance."* (Paris: DAC Working Party on Aid Evaluation).

Organization for Economic Cooperation and Development (OECD). 1992. *"Development Assistance Manual: DAC Principles for Effective Aid.* (Paris: Organization for Economic Cooperation and Development).

Participation. <online> available at: http://www.civicus.org. (Accessed on 12 June 2014). Prentice Hall.

Ramboll. 2005. Department of Water Affairs & Forestry, Republic of South Africa. *Project Monitoring and Evaluation*.

Raynor, Jared, Peter York and Shao-Chee Sim (2009). *What Makes an Effective Advocacy Organization?* A Framework for Determining Advocacy Capacity. TCC Group for the California Endowment.

Shapiro, J. 2001. *Monitoring and Evaluation*. CIVICUS: World Alliance for Citizen

Should Know. <Online> Available at http://datascientistinsights.com .accessed on 02 June 2014

StatSoft, Inc. (1997). *Electronic Statistics Textbook*. Tulsa, OK: StatSoft. WEB:

SWISS AGENCY FOR DEVELOPMENT AND COOPERATION (2000): External Evaluation: *Are we doing the right things? Are we doing things right?*

THE M AND E NEWS. <Online> available at: www.mande.co.uk/

UNDP (2002): Handbook on Monitoring and Evaluating for Results. <Online> available at: www.undp.org/eo/

UNDP, OESP (1997): *Who Are the Question-makers?* A Participatory Evaluation handbook. <Online> available at:intra.undp.org/ eo/methodology/methodology.html

UNFPA (2000): *Monitoring and Evaluation Methodologies: The Program Manager's M&E Toolkit.* <Online> available at: bbs.unfpa.org/ooe/me_methodologies.htm

UNICEF (1991): *A UNICEF Guide for Monitoring and Evaluation: Making a Difference?*

UNITED NATIONS CHILDREN'S FUND (UNICEF): *A Guide for Monitoring and Evaluation*<Online> available at: www.unicef.org/reseval/ mande4r.htm

UNITED STATES AGENCY FOR INTERNATIONAL DEVELOPMENT (USAID), CENTER FOR DEVELOPMENT INFORMATION AND EVALUATION (CDIE): Performance Monitoring and Evaluation Tips. <Online> available at: www.dec.org/usaid_eval/004

USAID, *Centre for Development Information and Evaluation.* <Online> available at: www.dec.org/usaid_eval/

W.K. KELLOGG FOUNDATION (1998): Evaluation Handbook. <Online> available at: www.WKKF.org

Weiss, N.A. 1999. *Introductory Statistics.* Addison Wesley.

WFP. 2014. *Monitoring & Evaluation Guidelines: Reporting on M&E Data and Information for EMOPs and PRROs.* Available at: www.wfp.org (Accessed on 3 July 2014)

WOODHILL J. AND ROBINS L. (1998): *Participatory Evaluation for Landcare and Catchment Groups: A Guide for Facilitators.* Australia

WORLD BANK (2001): *Impact Evaluation.* <Online> available at: www.worldbank.org/poverty/impact/index.htm

WORLD BANK (2002): *Monitoring and Evaluation Chapter* (draft), Monitoring and Evaluation for Poverty Reduction Strategies, www.worldbank.org/html/oed/evaluation/

WORLD BANK INSTITUTE: *Training Evaluation Toolkit* (Version 1.3). <Online> available at: www.worldbank.org/wbi/

WORLD BANK: *Evaluation, Monitoring and Quality Enhancement* Community Website. <Online> available at: worldbank.org/ html/ oed

WORTHEN, BLAINE R., JAMES R. SANDERS AND FITZPAT-RICK J. (1997): *Program Evaluation: Alternative Approaches and Practical Guidelines*, 2nd Edition, White Plains, NY

YRBSS. 2014. *Software for Analysis of YRBS Data.* <Online> available at: www.cdc.gov/yrbss. (Accessed on 2 July 2014)

Index

Coding, 87, 88, 89, 90, 91
Coding systems, 88
Conceptual Framework, 2, 78
Conceptualization, 78, 79
Confidence Interval, 62, 68, 69, 70
Confidence Level, 50, 76
Data
    Primary data, 2, 58
    Secondary data, 2, 30, 58
    Qualitative data, 59, 102
    Quantitative data, 58, 84, 85, 86
    Data organization
    Data matrix, 59
    Data input, 60
    Data analysis
    Data output, 60
Data Cleaning, 84, 86, 93
Data Collection
Data collection techniques, 13
Data Consolidation, 86
Data Management
Data Management Framework, 104, 106
Data Management Model, 77
Dummy variables, 88
Evaluation life cycle, 26, 29
Framework
Frequency graph, 47
Frequency table, 46, 47
Instrument design, 78, 80
Labeling variables, 85, 87
Logframe, 31, 32, 119, 123
Matrix
Mean

Median, 36, 49, 50, 62, 76
Missing value, 85, 86, 92, 93
Monitoring and Evaluation
Monitoring and Evaluation cycle
Monitoring and evaluation plan, 22
Monitoring and Evaluation report guidelines
Null hypothesis, 51, 52, 75, 77
Percentile, 46, 49, 50, 62
Pilot Testing, 78, 81
PRISM
    PRISM algorithm, 42
    PRISM criteria
    PRISM implementation
    PRISM outline
Qualitative data collection
Quantitative data collection
Report Writing, 33, 102
Research hypothesis, 51, 52
Research process, 11, 79
Research questions, 12
Sample size, 66, 84
Sampling
Sampling frame, 52
 Sampling methods, 53
    Purposive sampling, 54
     Simple random sampling, 64, 65, 66, 68
     Probability sampling, 13, 54, 84
     Clustered sampling, 54
Sampling unit, 52, 54, 59, 67
Scales, 55, 58, 59
    Nominal, 90, 93
     Ordinal, 82, 85, 89, 90, 91
     Interval
     Ratio

Self-assessment tool, 7
Significance Test, 51
Spider chart, 7
Statistical Software Packages, 60, 61, 62, 104
Structured questionnaire, 30, 94
Variables

Bongs Lainjo, MASc, Engineering

M and E: Data Management Systems

The author, Bongs Lainjo, MASc, Engineering is an RBM Systems Consultant with a bias to Program Evaluation. He is a former UN Senior Advisor, Program Management, Logistics and Evaluation. Before that, he worked for USAID as Logistics and Management Information Systems Advisor (LISA). Earlier, he served as COP/Senior Data Management Advisor for Columbia University after spending about a decade as professor in various Canadian Academic Institutions teaching in both English and French.

Professionally, Bongs' work has taken him to many countries including:

Uganda, Tanzania, Botswana, Swaziland, Zimbabwe, Ethiopia, Ghana, Nigeria, Lesotho, South Africa, Burkina Faso, Cameroon, Senegal, Niger, Kenya, Mozambique, Namibia, Maldives, Madagascar, Mauritius, Seychelles, Pacific Island Countries (PICs), Comoros, Nepal, Sri Lanka, India, Afghanistan, Bhutan, and the USA.

Some of his activities include developing and implementing thematic (peer-reviewed PRISM, RAPSYS etc) models; developing M and E systems; developing Household Survey protocols, RBM, M and E and Supply Chain Management System training manuals; conducting workshops on M and E, HMIS, Data Management and LMIS for UN staff and Program Implementing partners; as well as training national staff on conducting HIV/AIDs as well as house hold surveys. Further to this, he also conducted an M and E assignment in a USAID-funded project that included reviewing its Performance Management Plan (PMP) He has presented and continues to present papers at International conferences; and served as Front Line Advisor in Sri Lanka, the Maldives and Indonesia during the tsunami of 2004 and has served as a UN Security and CT member. He has also been an actively participating columnist to a largest newsmagazine in the city of Deerfield Beach, FL.